DREAM ISLANDS

THE WORLD'S MOST BREATHTAKING PLACES

Antony Mason

DREAM ISLANDS

THE WORLD'S MOST BREATHTAKING PLACES

ANTONY MASON

Quercus

82 Ellesmere Island
58 Spitsbergen
54 Lofoten Islands
38 Bornholm
20 Gotland
50 Surtsey
24 Isle of Skye
28 Staffa
30 Rathlin Island
12 Venice
78 Mont-St-Michel
62 Hvar
70 Ile de Ré
108 Nantucket, Martha's Vineyard
74 Corsica
8 Capri
112 Florida Keys
34 Madeira
16 Sicily
92 St Barthélemy
116 Hawaii
86 Cuba
98 Guadeloupe
66 Kornati Islands
94 Martinique
90 Aruba
40 Kefalonia
102 St Lucia
44 Santorini
126 Galápagos
46 Crete
134 Bora Bora
130 Tahiti
122 Easter Island
104 Tierra del Fuego

CONTENTS

172 Hokkaido
176 Ryukyu Islands
178 Hainan
160 Borneo
164 New Guinea
120 New Georgia
156 Phuket
140 Maldives
144 Seychelles
168 Bali
186 Fraser Island
148 Mascarene Islands
152 Madagascar
138 New Caledonia
182 Tasmania

INTRODUCTION

What constitutes a dream island? A tropical coral atoll may spring first to mind – white-sand beaches, shaded by coconut palms rustling gently in the cooling sea breezes, with a view out over limpid seas vanishing in gradated blues of every nameable hue – turquoise, aquamarine, cobalt – towards a distant horizon suspended beneath spindrift clouds.

The idea of dream islands seems to be part of our psychic make-up. They appear in Ancient Greek myths, and in the sagas of the Vikings. Each consecutive era has its own version. The coral-atoll dream is a modern one: beach bathing was considered an eccentric pursuit before 19th-century physicians promoted the therapeutic benefits of saltwater and fresh air. And sunbathing became fashionable only in the 1920s – supposedly after Coco Chanel returned from holiday in the South of France with an accidental tan.

The experiences of US forces in the South Pacific during the Second World War brought many ordinary Americans into contact with atoll islands for the first time. James A. Michener's *Tales of the South Pacific* (1946), based on his war experiences, were turned by Hollywood – the world's greatest purveyor of dreams – into the blockbuster musical *South Pacific* in 1958, just as long-distance air travel was about to become cheaper and quicker. Suddenly islands such as Mauritius, the Seychelles and Aruba, which had struggled to live off their own natural resources and plantation produce, found they were blessed with a combination of assets that the world would pay top money for: sun and sand.

Be it in the South Pacific or cold northern climes, there is something essentially comforting about islands. Standing on an island and looking back at the mainland can incite a strange, tingling pleasure: so close, but one step removed. On the Ile de Ré residents refer to the rest of France – which is in sight along the entire northeast coast – as *Le Continent*, almost dismissively, as though they are not a part of it, and rather like it that way. Similarly, in Martha's Vineyard and Nantucket, the mainland is referred to as America: not us, but them. By contrast, totally isolated islands, surrounded by nothing but sea stretching to the distant horizon, can offer the liberation of utter disconnectedness – a mental and physical framework to test just what resources one has or in fact needs. This can reinforce a healthy sense of self-sufficiency, and maybe even redefine one's identity as a human being.

In the distant past, islands were cherished because they provided refuge: a dream island in Scotland or Ireland during the times of Viking raids had good natural defences. Enemies are exposed as they approach by water, giving islanders at least some chance of not being taken by surprise. Humans are not alone in appreciating this aspect of islands: it is clearly a natural instinct. Sea birds, for instance, will choose islands as nesting sites – and in the case of Rathlin Island in Northern Ireland, in their hundreds of thousands.

Islands that have been isolated geographically for thousands of years develop their own unique species of plants and animals. This is true of Madagascar, with its lemurs and baobab trees, and New Caledonia similarly has a broad range of its own indigenous plants and animals. Most famously, the highly individual wildlife of the Galápagos Islands – with its giant tortoises, marine iguanas, and species of finches that differ from island to island – provided the inspiration for Charles Darwin's theory of evolution. Without the laboratory conditions of isolated islands, this line of thinking would have been much less persuasive and may not have surfaced at all.

The same goes for human populations to a degree, and on their much shorter historical timescale. Because of their natural isolation, islands reinforce an individualized sense of community: all islands feel a little different – different from the mainland, different from each other, even when close together. Very isolated islands can foster quite distinct cultures over a matter of centuries. Perhaps the most celebrated example of this is Easter Island, where Polynesian settlers evolved their unique tradition of carving massive stone heads.

Easter Island is an extreme case: many less isolated islands have also developed unique cultures. Venice's lagoon protected it from its enemies, and also allowed it to create a city on water which, if not the only one, is by far the most sophisticated. Bali was able to preserve its Hindu traditions only by virtue of being an island, and by closing its doors to almost all incomers for centuries until the final Dutch invasion in 1906.

And that is the downside of islands: while the isolation of islands may provide some sense of security and refuge, it also makes them vulnerable to covetous and determined enemies. The histories of the islands of the Mediterranean, big and small – from Corsica and Sicily to Santorini and Kefalonia – are catalogues of competing nations and empires: Greek, Roman, Byzantine, Arab, Venetian, Ottoman, French, British, and with a fair dose of piracy along the way.

It is tempting to gloss over the horror that lies behind some of the places that now qualify as dream islands. On idyllic Greek islands, mothers would choose to scar the faces of their daughters to try to prevent them from being kidnapped and enslaved by marauding pirates. The Caribbean islands were attractive to European settlers because fortunes could be made out of sugar – but sugar cane demanded a big labour force: this could be conveniently and cheaply supplied by African slaves, and by plantation owners turning a blind moral eye to all that this entailed. Even after slavery had been outlawed, 'blackbirders' haunted the islands of the South Pacific to capture and ship off vulnerable inhabitants to plantation islands. Often, too, indentured labour from India, Malaya and China, on minimal wages and with punitive working contracts, were recruited to fill the gap in labour left by the end of slavery. Today, we can make a virtue of the blend of cultures that this history produced on many islands – but it came at a considerable cost.

Such outcomes were by no means inevitable: dream islands were always supposed to be ideal places. This was a vision that inspired European explorers as they sailed forth in the Great Age of Exploration of the 15th and 16th centuries, in search of Cathay, or Atlantis, or King Solomon's mines. When Sir Thomas More envisaged the perfect world of Utopia in his book of 1516, he placed it on a fictional island of that name in the Atlantic. The harsh reality of exploration and human greed quickly put paid to such fantasies, but the concept of dream islands never entirely vanished. It returned to centre stage in the 18th century when philosophers debated the essential nature of humans beings. Were they naturally virtuous, or could virtue only be taught? Islands could serve as the proving grounds.

When Daniel Defoe published *Robinson Crusoe* in 1719, he explored what could happen to a person in total isolation on a desert island: the book has captivated audiences ever since, right around the world, and each successive generation has its own interpretation, flavoured by its era. The French writer Jacques-Henri Bernardin de St-Pierre used French-ruled Mauritius as a setting for his romantic, and ultimately tragic, novel *Paul et Virginie*, first published in 1787: with the questions it poses about human ideals, and how we might construct a world to match them, it captivated French readers on the threshold of the French Revolution.

By this time European ships were visiting the islands of Polynesia, notably Tahiti, where their crews were at once intrigued and confused by the contentment of islanders leading technologically unsophisticated lives. As Herman Melville described in his first published work *Typee: A Peep at Polynesian Life* (1846), a novelized account of his own real experience, the allure of island life, and the uncomplicated sexuality of the women, was too great for many visiting sailors, on whaling ships as much as naval vessels: risking savage punishment at the hands of search parties, they jumped ship.

The result was usually disaster. Dream islands are, by definition, fantasies, and too often undermined the deadening truth that 'familiarity breeds contempt'. But on many of the dream islands in this book people have chosen to live out their fantasies and have not looked back – artists in Bornholm, writers on Caribbean islands, diving professionals in the South Pacific. The prerequisites of a dream island are as different and various as islands themselves; the seas contain as many of them as there are human imaginations.

Antony Mason

CAPRI

Latitude 40°33'N **Longitude** 14°14'E

Area 10.4 square kilometres (4.02 sq miles)

Status Island of the region of Campania, Italy

Population 12,200

Capital Capri town

Official language Italian

Currency Euro

Capri has enchanted visitors for more than 2,000 years. A rocky full point to the exclamation mark of the Sorrento Peninsula, lying 7 kilometres (4.3 miles) off the mainland, it looks out into the Tyrrhenian Sea on one side, and, on the other, commands entrancing views across the Bay of Naples to the brooding contours of Vesuvius. The landscape of Capri itself is beguiling, a raised saddle of limestone, greened by pines, cypresses, myrtle, juniper, acanthus and vines. Limestone rims the island, creating a coastline of contorted and honeycombed cliffs, with few beaches.

Offshore, a group of three limestone seastacks called I Faraglioni are a famous landmark.

Visitors arrive by ferry at the Marina Grande from Naples or ports on the Sorrento Peninsula. A funicular train takes them up to the main town, also called Capri, which stands 150 metres (500 ft) above sea level. The winding streets, domed churches and intimate central square speak of the town's long history, while chic boutiques and jewellers are reminders that Capri has – and always did have – a reputation for luxury.

Mount Solaro is the highest point on Capri [589 metres/1,930 ft], offering spectacular views across the Bay of Salerno.

TRAVELLER'S **TIPS**

Best time to go: Any time of year is good. During the summer months (June-September) the island becomes very crowded.

Look out for: A chairlift takes visitors from Anacapri to the top of Monte Solaro, the highest point of the island, with superb views all round.

Dos and don'ts: Do visit the Blue Grotto, but choose a day when the weather is calm (for the boat ride) and the sun is bright (for the light).

Inhabited since Neolithic times, the island was settled by the Greeks, who established trading colonies and towns all along this coast. Capri town has a counterpart, and former rival, called Anacapri on the western end of the island. Separated by a ridge, the only access to it by land – until a road was built in 1874 – was by means of the Scala Fenicia, a flight of nearly 900 steps cut in Greco-Roman times. The Romans were the first to make the island into an exclusive resort. The emperors built 12 villas here. The extensive ruins of Villa Jovis, on a high ridge at the eastern end of the island, hint at the vast complex that once stood here. This is where the Emperor Tiberius (reigned AD 14–37) lived out his last ten years, secure in reclusive isolation, signalling commands across the Bay of Naples using light and smoke. Following the Roman era, Capri came under control of the Byzantines, who built chapels and monasteries here. It was frequently raided by Saracen pirates, and in 1535 it was captured by the Ottoman admiral Barbarossa. During the Napoleonic wars it was the turn of the French, then the British to seize the island.

In 1826 a German poet and painter called August Kopisch saw and publicized the Grotta Azzurra (Blue Grotto). This large sea cave has an entrance so small that visitors, taken by guides in rowing boats, have to lie back in the boat to enter. Inside, light enters from another larger submarine hole, turning the water a sensational, iridescent blue. The Swedish physician Axel Munthe (1857–1949) owned the beautiful Villa San Michele; when he published his bestselling memoir *The Story of San Michele* in 1929, he convinced new audiences about the charms of Capri. Meanwhile signed photographs of filmstars and other celebrities hanging in restaurants testify to the glamour that has always been associated with the *Isola dei Sogni* – the Isle of Dreams.

On the south side of the island is the Grotta Verde (Green Grotto), less well known than the Blue Grotto on the north side, but comparably beautiful. Colours here vary from turquoise to emerald green, as sunlight bounces off the white-sand floor – best enjoyed by swimming right into the cave.

The Marina Grande is the main port, where ferries arrive from Naples, Sorrento and Amalfi, sharing the harbour with pleasure craft. This first sight of Capri reveals the island's dramatic limestone skyline.

VENICE

Latitude 45°26'N **Longitude** 12°20'E

Area 415 square kilometres (160 sq miles)

Status Italian city, capital of the Veneto region

Population 270,600

Capital Venice

Official language Italian

Currency Euro

Any city in the world with canals and waterways is casually compared to Venice, but nowhere comes near to matching Venice itself. For one thing, Venice is an island, or rather a cluster of more than 100 islands, set in the midst of a large lagoon. It has evolved in a unique way by sticking to the original concept: that boats would always be the main means of transport, and canals the main thoroughfares – a rule that still applies to the traditional, oar-powered gondolas and the modern *vaporetti* water buses. Add to this that Venice invested its great wealth – acquired through its medieval trading empire – in the

Gondolas at dawn, moored outside the Palazzo Ducale, face across Canale di San Marco towards the church of San Giorgio Maggiore, which stands on an island of its own.

TRAVELLER'S **TIPS**

Best time to go: May–June and September have the best weather and light. July–August is the busiest time. Carnival takes place at the start of Lent.

Look out for: The Scuole – headquarters of historic charitable confraternities decorated by the great masters such as Tintoretto and Tiepolo.

Dos and don'ts: Do get off the beaten track. Wandering around the network of canals away from the main tourist areas reveals the true character of the city.

finest art and architecture, and we arrive at a floating dream city that has entranced visitors for more than five centuries.

Enchantment lies at every turn: in the grandeur of St Mark's Square with the Byzantine domes and gilding of the Basilica of St Mark, and the soaring Campanile (belltower); in the elegant *palazzi* that line the serpentine Grand Canal; and also in the labyrinths of alleyways, backwaters, bridges and squares that run through the interlocking islands. The historic wealth, self-esteem and magnificence of the city can be witnessed in the richly frescoed halls that lie behind the filigree façades of the 14th-century Palazzo Ducale, headquarters of the Doge, elected leader of the city. The Galleria dell'Accademia shows the ability of the city to nurture outstanding artists over the centuries, such as Giovanni Bellini, Giorgione, Titian, Tintoretto, Veronese, Tiepolo, Canaletto and Guardi.

And, remarkably, Venice remains today more or less as Canaletto painted it almost three centuries ago. Venice's

story began at the fall of the Roman Empire, when people of the Veneto region took refuge from invading Huns on marshy islands in the lagoon. To build on the islands, they drove wooden piles deep into the mud, and bit by bit created a city, first in wood and then in stone. Having no agricultural land, the Venetians had to trade: they headed into the eastern Mediterranean, cornering the market for precious goods arriving from Asia and along the Silk Road from China – spices, incense, porcelain, silk, gemstones.

Venice's empire grew, with outposts and colonies across the entire Mediterranean. It tussled with the other growing force of the region, the Ottoman Empire, confronting it with its formidable navy of galleys. Venice's dockyard and munitions factory – the Arsenale – was reputedly able to produce a new warship in just 12 hours.

When Vasco da Gama, the Portuguese navigator, rounded the Cape of Good Hope in 1498 and headed to India, he blazed the trail of European trade and exploration in the Far East that would undermine Venice's monopoly. La Serenissima ('Most Serene Republic'), as Venice liked to call itself, declined in power, but nonetheless retained its reputation for high art, music and luxury, as well as decadence. The four-month-long

masked Carnival season was notorious for its licentiousness, and attracted the nobility of Europe on their Grand Tour.

Napoleon put a stop to that when he seized the city in 1797. The city changed further when, in 1846, a railway bridge bound it to the mainland for the first time. In 1966 disastrous floods warned that Venice was slowly sinking into the mud. To stop this, the drilling of artesian wells on the edge of the lagoon was banned. Now, with the installation of experimental inflatable pontoons designed to control the rising flood levels in the lagoon there is cautious optimism that La Serenissima is not, after all, doomed.

Four bronze horses stand on the façade of St Mark's Basilica, overlooking the Piazza San Marco and the Campanile (belltower). These are copies; the original Ancient Greek horses (now in the Basilica Museum) were used by Roman Emperor Constantine in the 4th century to decorate a race track in Constantinople.

Lined by some 200 elegant *palazzi* (palaces) built between the 12th and 18th centuries, the Grand Canal snakes its way for nearly 3 kilometres (2 miles) through the heart of Venice, providing access to a web of side canals.

SICILY

Latitude 38°6'N **Longitude** 13°21'E

Area 25,711 square kilometres (9,927 sq miles)

Status Autonomous region of Italy

Population 5,050,000

Capital Palermo

Official language Italian

Currency Euro

The Scala dei Turchi ('The Turks' Staircase') is a natural wonder on the south coast near Agrigento. Uplifted layers of chalky marl have been eroded into a set of huge, curving, bright-white steps.

When Archimedes, the Ancient Greek mathematician and engineer, ran down the street naked shouting 'Eureka' in triumph after making his discovery about the displacement of water in his bath, it was in the city of Syracuse, Sicily. It is all part of the rich and complex history of the island, which conspires to make it quite unlike any other place. Although a part of Italy since 1861, and separated from it by just 3 kilometres (2 miles) at the Strait of Messina, Sicily has a quite distinct identity. The language Sicilian, still widely spoken, is merely a cousin of Italian, with loan words from Greek, Arabic, Catalan, Spanish and French – an indicator of the strange mosaic of history that has shaped the island.

TRAVELLER'S TIPS

Best time to go: Spring (March–April) is celebrated for its wild flowers and agreeable temperatures. High summer can be very hot.

Look out for: The Sicilian desserts of *cassata*, an ice-cream cake (from Palermo); *granita*, semi-frozen sorbet (Catania); and *cannoli*, creamy pastry tubes (Noto).

Dos and don'ts: If on a driving tour, don't be too ambitious: the roads in the mountainous interior and on the coast can twist and turn, and progress is slow.

Sicily is the largest island of the Mediterranean, and hangs off the toe of the boot-shape of the Italian mainland rather like a football – if a triangular one. Inland it has a rugged and mountainous terrain, much of it turned over to agriculture, producing oranges and lemons, olive oil and its distinctive wines – rich, red and spicy, and redolent of the intense summer heat. Fishing also plays a major role in the economy, and the coast has plenty of sandy beaches that make this a popular destination for summer holidays. Visitors come also to see the great cultural riches of Sicily's 20,000-year human history. Strategically placed in the middle of the Mediterranean, it has always been a prized trading hub. Greeks settled here sometime after the fall of Troy in around 1200 BC: the sea monsters Scylla and Charybdis in Homer's *Odyssey* are taken to be a reference to the Strait of Messina. Some of the world's best-preserved Greek temples are found in Sicily, notably in the Valley of the Temples near Agrigento.

The Romans destroyed the Sicilian Greeks, who had mistakenly sided with Hannibal and the Carthaginians in the 3rd century BC. Sicily became a rural backwater – but the Romans left some outstanding remains, including the mosaics of the Villa Romana del Casale near Piazza Armerina – famous for its pictures of women gymnasts in bikinis. In the 11th century the Norman adventurer Roger Guiscard took the island and launched a golden age that lasted 100 years, during which Christians co-operated with Arab settlers, and beautiful churches were built, glittering with mosaics, such as the 12th-century cathedral of Monreale.

Mount Etna looms over the Gulf of Catania, dominating eastern Sicily. It is the highest volcano in Europe, rising to 3,320 metres (10,890 ft) – but the height changes because Etna is still active. Since it erupts regularly, the internal pressure is released, so destruction tends to be limited.

It has been Sicily's fate to be governed by a succession of foreign dynasties. So in the 13th century it was ruled by German Hohenstaufens, then by Charles of Anjou (France), the Kings of Aragon and Princes of Castile (Spain), the Duke of Savoy, the Emperor of Austria and the Spanish Bourbon kings of Naples. It has suffered numerous wars and civil wars, which explains its many castles and hill-top towns: strikingly beautiful today, they have their origins in practical necessity.

This perhaps also explains why Sicily has a reputation for secretiveness, and for traditions of clan loyalty, protectionism, honour and vendetta that translated to the criminal syndicates of the Mafia, still so readily associated with the island. But, while this problem has not gone away, Sicily has cleaned up considerably, making it more than ever an island of dreams.

The Cathedral of San Giorgio, Modica, southern Sicily, is typical of the Baroque style that took hold of Sicily in widespread rebuilding following an earthquake in 1693. With extravagant flourishes layered onto a neoclassical base, this was the triumphant style of the Catholic Counter-Reformation. A staircase of 250 steps leads up to the entrance.

GOTLAND

Latitude 57°30'N **Longitude** 18°33'E

Area 3,140 square kilometres (1,212 sq miles)

Status Province of Sweden

Population 58,000

Main town Visby

Official language Swedish

Currency Swedish krona

Gotland is famous for its laid-back pace. This is a place to unwind, to retreat from the pressures of modern life. One super-stylish boutique hotel, with a top-quality restaurant, also offers simple hermit's cabins set in coastal woodland for guests who really want to get away from it all: tiny, cosy little shacks deliberately deprived of electricity and running water.

Gotland is, and always has been, a place apart. Of Sweden's 221,800 islands (only 401 of which are inhabited) it is by far the largest. It sits in the Baltic Sea, with the Swedish mainland

TRAVELLER'S **TIPS**

Best time to go: The summer high season is mid-June to mid-July, and the busiest time. But visitors can enjoy Gotland at any time of the year.

Look out for: Traditional games such as *kubb* (sticks thrown at wooden blocks), *pärk* (a ball game) and *varpa* (resembles boules).

Dos and don'ts: Do visit the Bungemuseet, an open-air museum with a collection of 50 rescued rural buildings (some furnished) from the 16th to 19th centuries.

about 80 kilometres (50 miles) to its west, and 160 kilometres (100 miles) south of Stockholm. The coast of Latvia lies 150 kilometres (93 miles) to the east. Visby aside, Gotland has always has been primarily rural, with picturesque farmhouses and painted cottages sitting among a low-lying, gently undulating landscape of conifer woodland and fields. It is celebrated for its wildflowers – bright blue viper's bugloss by the roadsides, poppies nodding among the ripening wheat, and hosts of wild orchids. Gotland also has a strong sense of its unique identity. The island has been inhabited since

St Mary's Cathedral rises above the low medieval skyline of Visby, and its network of crooked streets lined with pastel-shaded and half-timbered houses. The 13th-century cathedral is among some 200 buildings that recall Visby's Hanseatic days, and the city is ringed by one of the best-preserved sets of medieval defensive walls in Europe.

Old fishing huts at Helgumannen, Fårö, part of the Gotland group, are a potent reminder of the islands' close bond with the sea.

prehistoric times. Another name for the Gotlanders is Gutes (and their name for the island is Gutland): according to the old sagas that speak of the early history of the island, the Gutes ventured far and wide as seafarers and traders. Some, it seems, travelled down the trading corridor of east European rivers to reach the Black Sea and settled on the peripheries of the Greek world. They may well have been the original Goths – one of the 'barbarian' groups that harried, destroyed and eventually took over the Western Roman Empire.

The Gotlanders remained vigorous rovers and traders throughout the later Viking era, recalled in numerous burial sites, jewellery and treasure hoards, and memorial stones inscribed with vivid pictures and intricate patterns. But they reached their true golden age in the 13th and 14th centuries, when Gotland became the most important Hanseatic trading centre of the Baltic Sea. Visby was now one of the wealthiest and most sophisticated cities of northern Europe. Many of the 94 historic parish churches of Gotland were built on the back of this wealth, using the latest, fashionable styles of continental Europe: first Romanesque, then Gothic. Silver Arab dirhams have been found in treasure troves on the island, a testimony to the reach of Gotland's trade; they, and the Viking treasures, can be seen today in the Fornsalen museum in Visby.

Then came the Danish conquest of 1361 – a date of doom in Gotland folklore. The island now went through turbulent times, as it was invaded and ruled by privateers, then Teutonic Knights and then in 1409 was sold to the Danish crown; it was finally assigned by treaty to Sweden in 1645. As Gotland's trading power melted away, the island turned into a kind of peasant republic, run at parish level, and only loosely

controlled by the Swedish authorities. During the 17th, 18th and 19th centuries, in self-imposed semi-isolation, the Gotlanders reinforced their folklore traditions: music, dance, costumes, poetry and song. This proud sense of cultural identity remains a key feature of the island today, and has proved a magnet for artists, musicians and filmmakers (Ingmar Bergman lived and filmed on the island of Fårö, just to the north of the main island).

Heritage is a major part of Gotland's appeal. But it is also a summer holiday destination, with excellent sandy beaches, and a vibrant youth scene. It receives some 1.5 million visitors a year, most of them in the summer season. Medieval Week in August is a high point: it is a festival in and around Visby, when Gotland's heyday is evoked and recalled through jousting, markets, street entertainment, music and dance.

Erosion by wind, sand and sea has carved bizarre sculptural shapes, called *rauk*, out of limestone on the eastern beaches, as here on Fårö. Some are pillars, some arches and some defy categorization, inspiring names such as 'The Coffee Pot' and 'The Old Man of Hoburg' – referring to legendary 'stone trolls' who turned into stone during the day.

Gotland is celebrated for its beaches, typically lined with salt-tolerant marram grass. Some are pebbly, such as Ireviken on the north coast, while the silky-soft sand at Tofta Strand, on the west coast, can draw huge crowds on a sunny summer's day.

ISLE OF SKYE

Latitude 57°18'N **Longitude** 6°13'W

Area 1,656 square kilometres (639 sq miles)

Status Island of the Inner Hebrides of Scotland

Population 9,230

Main settlement Portree

Official language English (and Gaelic)

Currency Pound sterling

An t'Eilean Sgitheanach is its name in Gaelic, the historic language of the Isle of Skye. The meaning 'winged island' is disputed, but it is certainly appropriate. The ragged coastline, fretted by deep, fjord-like inlets or lochs, creates a shape reminiscent of the outstretched wing of a great bird, like one of the golden eagles that soar over the sea-cliffs here.

And then, of course, there are echoes in the 'The Skye Boat Song', one of Britain's most haunting ballads: 'Speed, bonnie boat, like a bird on the wing… Over the sea to Skye'. It tells the story of Bonnie Prince Charlie's dramatic escape to Skye, aided by the beloved local heroine Flora MacDonald, after the calamity of Battle of Culloden in 1746; he later sailed to France, carrying the hopes of Jacobite rebels with him.

The island is punctuated with memories of that story, some real, many fanciful, attached to homesteads, rocks, caves and graveyards. Dotted all over the island, too, are mementoes of a long and deep, and often rough history. A 2,000-year-old broch (round tower) at Dun Beag, near Bracadale, served as a defensive refuge of stone, where farmers and their livestock could hold out against marauders. Vikings, who attacked, then settled and ruled the island, have left a legacy of Norse place names: Skye, the largest island of the Inner Hebrides, with swathes of fertile land, was a valuable prize. And then there are the many ruins of castles of the rival clans, notably the MacDonalds, at Knock Bay, Dunsgiath and Duntulm, for instance. The impressive surviving castle at Dunvegan, the seat of the MacLeod clan since at least 1200, has been softened by Georgian and Victorian country-house upgrades, but still has the gaunt, muscular exterior of a medieval fortress.

Portree, on the east coast, is the only town of any size on Skye. Pastel-painted houses on Quay Street look out over a sheltered natural harbour.

TRAVELLER'S **TIPS**

Best time to go: The summer months have long days, often lit by superb (if changeable) weather. But always be prepared for rain.

Look out for: The peaty single malts produced by Talisker; you can visit the distillery, set close to the Cuillins.

Dos and don'ts: Take some insect repellent: on fine, still days in summer (May–September) biting midges can be a real nuisance.

The Old Man of Storr is a noted landmark. The largest in a group of giant basaltic pillars, it rises to 49 metres (160 ft). Views to the east lead the eye over Loch Leathan and the Isle of Raasay to the Scottish mainland.

Life was never easy on these islands. Crofters eked a living from sheep farming and cultivation, living in simple cottages of stone and thatch which are known as 'black houses'; this history is sensitively revealed at the Skye Museum of Island Life, near Kilmuir in the north of the island.

Skye once had a population numbering tens of thousands, but the Highland Clearances in the mid-19th century – evictions by landlords to make agricultural land more economically viable – forced some 30,000 islanders to emigrate. A famous revolt on Skye against Lord MacDonald ended up with the Battle of Braes in 1882, which helped shame the government into land reform. Skye remains lightly populated to this day: the only town is Portree, set around its picturesque harbour.

All this history speaks of the struggle to try to come to terms with the one dominant and overriding feature of Skye: landscape. Humans often seem barely relevant. Even the swelling numbers of tourist visitors, who cross the controversial bridge that has anchored Skye to the mainland at the Kyle of Lochalsh since 1996, fail to upset the equation.

The south is dominated by the awe-inspiring Cuillin Mountains, which soar from the coast to 992 metres (3,255 ft), sometimes brooding beneath thunderous clouds, sometimes swathed in mist (Skye is also known as Eilean a' Cheò, 'Island of the Mist') and sometimes serenely silhouetted against a brilliant blue sky. The views of the Cuillins across the southern lochs can be majestic. Three of the Munros are here – the 283 Scottish peaks over 914 metres (3,000 ft) that throw down the gauntlet to the most determined hillwalkers. On the 'Inaccessible Pinnacle' of Sgurr Dearg they face the Munros' greatest challenge: the only peak that requires full rock climbing.

On the Trotternish peninsula in the northeast, a ridge of basalt has created the outlandish landscape of the Quiraing, peopled by giant pillars of rock. A little to the south, the giant basaltic pillar called the Old Man of Storr rises 49 metres (160 ft) from its base and cuts a lonely figure against a wild landscape often veiled in drizzle and swirling mists.

Close to the coast below the Old Man of Storr a tenth-century treasure hoard was discovered in 1891. The contents, which are now in the National Museum of Scotland in Edinburgh, included precious brooches, bracelets and clasps and gold coins, some of them from Samarkand – a startling reminder of the Viking trade routes that once stretched right across Europe to the Middle East. It is a reminder, too, that, if the beautiful and defiant landscape of Skye is linked to the great scope of human history, it will always be on its own terms.

STAFFA

Latitude 56°26'N **Longitude** 6°20'W

Area 0.34 square kilometres (0.13 sq miles)

Status Island of the Inner Hebrides of Scotland

Population 0

Capital None

Official language English

Currency Pound sterling

Just off the west coast of Mull lies a wonder of nature. Sandwiched into the rocky little uninhabited island of Staffa is a thick filling of near-perfect hexagonal columns of basalt, standing some 20 metres (65 ft) tall, tightly packed, neatly topped and tailed, in gently undulating contours.

On the south side of the island, erosion has a carved a sea cave, where the columns line up like the massed pillars of a cathedral. It is called Fingal's Cave, so named because it seems so unreal: surely this is not the hand of nature, but of some super-human force? Fingal (or Fionn Mac Cumhaill) was a mythical Irish hero of monstrous proportions and bravery. His lady-love lived on Staffa, so Fingal built a magnificent path for her across the sea to Ireland, arriving at a similar set of unearthly hexagonal columns on the coast of Co. Antrim, 120 kilometres (75 miles) to the south, called the Giant's Causeway.

The columns were created 55–58 million years ago by a rare volcanic event. A mass of unusually fine-grained basalt lava formed into a thick layer then cooled at a consistent speed, putting an evenly distributed force of contraction on the crystalline structure of the rock, which then cracked into neat multi-sided, mainly hexagonal, columns. Sir Joseph Banks, famed as Captain Cook's botanist, came to Staffa in 1772 and marvelled at it. His reports caught the public's attention, and a stream of visitors began to take boats from the little ports on neighbouring islands – as they still can – to explore Staffa and wonder at its strangeness. It fitted perfectly the admiration of the Romantics for the raw power and limitless invention of nature. Sir Walter Scott, William Wordsworth, John Keats, Jules Verne and Queen Victoria all came here and were filled with a wonderment that remains as powerful to this day.

Many people know Fingal's Cave from Mendelssohn's *Hebrides Overture* (also known as *Fingal's Cave*), which he wrote after an 1829 visit to Staffa.

TRAVELLER'S **TIPS**

Best time to go: Boat trips are available April–September, but only when the sea is sufficiently calm. The weather here is notoriously fickle.

Look out for: The island supports a large number of nesting seabirds, including puffins. You may also see grey seals and dolphins, even minke and pilot whales.

Dos and don'ts: Some boat trips view Fingal's Cave from the sea; others land and allow passengers to explore on foot. Choose the one you want.

RATHLIN ISLAND

Latitude 55°17'N **Longitude** 6°11'W

Area 13 square kilometres (5 sq miles)

Status Island of Northern Ireland, UK

Population 70

Main settlement Rathlin Harbour

Official language English

Currency Pound sterling

Thousands upon thousands of birds. That is what the visitors come to see, ploughing their way on the open deck of a small ferry through the choppy waters and spray of the 10 kilometres (6 miles) that separate Rathlin from Ballycastle, on the north coast of Northern Ireland.

Guillemots, puffins, kittiwakes, razorbills, fulmars – from May to July they crowd and clamour and swirl around the shelves and turf summits of the cliffs and guano-coated sea stacks on the west coast of the island. The UK's Royal Society for the

A view from White Park Bay, Co. Antrim, shows the undulating contours of Rathlin Island, on the other side of Rathlin Sound.

Protection of Birds (RSPB) has a staffed Seabird Centre here, perched halfway up a cliff at the old lighthouse called the West Light.

The Rathlin Island Ferry reaches the calm of the broad crescent-shaped harbour, which lies sheltered behind two projecting moles, with the scoop of old volcanic hills to its back. Seals come here to bask. From the harbour, visitors will spread out across the island, on foot, or on hired bicycles, or by minibus. Those who head right from the harbour will reach

TRAVELLER'S **TIPS**

Best time to go: May to the end of July is the nesting season for seabirds, and the peak time for visitors; the summer season lasts until September.

Look out for: Puffins – Britain's most appealing seabird. They nest on the island from spring to early August. Early June is the best time to see them.

Dos and don'ts: Do book your ferry tickets in advance in the high season, especially at weekends between May and September. Don't forget binoculars.

the East Lighthouse, site of the cave where in 1306 Robert the Bruce, King of Scotland, is said to have watched a spider make its web. On the run from the armies of Edward I of England, he had crossed the channel to Rathlin from Scotland – a distance of just 24 kilometres (15 miles) from the tip of the Mull of Kintyre – to take refuge with the Bissett family. After many failed attempts, the spider succeeded in constructing its web, thereby teaching Robert the virtues of patient persistence, which he put into practice on his path to victory at the Battle of Bannockburn in 1314.

L-shaped like an upturned boomerang, Rathlin is the most northerly point of the northeast coast of Ireland, in the narrow gateway to the Irish Sea. In the past a population of a thousand could survive here, but the island's position has always made it a strategic target. This is where the first Viking raid in Ireland is said have struck, in AD 795, pillaging and burning the early Christian settlement founded by St Columba and St Comgall two centuries earlier. Extreme violence was a feature of repeated incursions. After Bannockburn English sympathizers in Ulster took revenge

on the Bissetts by dispossessing them. The MacDonnells took over in the 16th century, but in 1575 English forces massacred the islanders when they came to oust their leader, Sorley Boy MacDonnell. In 1642 the Campbells, arch-rivals of the MacDonalds in Scotland, again attacked, and threw dozens of women and children over the cliffs to the deaths.

Such outrages are hard to imagine in the deep serenity of Rathlin today. In fine weather, the captivating views across the sculptured coastline back to the mainland underline the island's self-contained separateness. Guglielmo Marconi felt that it resembled a ship at sea, and in 1898 set up colleagues on the island to test the possibilities for shore-to-ship signalling by radio: huge masts erected here and at Ballycastle formed the world's first commercial wireless telegraphy link.

In 1987 millionaire businessman Richard Branson again placed Rathlin in the record books, landing in the sea here at the end of the world's first ever transatlantic flight in a hot-air balloon. Homing signals are a part of Rathlin's natural history.

Bell heather and yellow dwarf gorse bring summer colour to the pasture above the cliffs of Altachulle Bay on the north coast. The drystone walling is a legacy of the island's history of sheep and cattle farming.

MADEIRA

Latitude 32°39'N **Longitude** 16°54'W

Area 801 square kilometres (309 sq miles)

Status Self-governing region of Portugal

Population 247,400

Capital Funchal

Official language Portuguese

Currency Euro

Madeira was first inhabited in 1420 – a measure of its remoteness. Way beyond the shipping lanes of the ancient world, it sits out in the Atlantic some 900 kilometres (560 miles) from Portugal, the country that laid claim to it, and on the same latitude as North Africa. The island is just the final crest of a vast submarine mountain that begins 6 kilometres (3.7 miles) beneath the surface of the Atlantic, thrust up by volcanic activity that began 4.6 million years ago and ceased in around 4500 BC. What remains is a dramatic landscape of steep, heavily eroded hills, rising to 1,862 metres

The knobbly hills near Porto da Cruz, in the northeast of the island, bear witness to the volcanic contortions of Madeira's origins. The massive rock formation in the middle distance is the Penha de Águia ('Eagle Cliff'), which rises 600 metres (1,970 ft) from the sea.

(6,108 ft) at Pico Ruivo. In 1418, at the very start of the Portuguese quest to find the Far East, the caravel of João Gonçalves Zarco was blown off course and onto Porto Santo, a small island to the northeast of Madeira. The following year, Zarco was despatched to explore further, and he ended up on Madeira. At that time, the island was heavily forested: hence its name Ilha da Madeira: 'Island of Wood'.

Settlers from Portugal soon began tearing back forest to grow crops. Since the northwest of the island receives more rain than the southeast, they constructed ingenious aqueducts, *levadas*, to distribute the water. Channels were cut into cliff-sides, and long tunnels dug through mountains. When the network ceased growing in the 1940s, its total length ran to 2,170 kilometres (1,350 miles). Paths beside these *levadas* provide the routes for many of the most exhilarating walks on the island today.

Because climatic conditions change with altitude, Madeira can produce an extraordinary variety of crops. Papayas, bananas,

TRAVELLER'S **TIPS**

Best time to go: The weather is good all year, with average temperatures of 26°C (79°F) in August. September–October and March–April are rainiest.

Look out for: The views from the island's highest point, Pico Ruivo, are spectacular. The peak is just a 40-minute walk from the nearest car park.

Dos and don'ts: Do make sure you are fully prepared when you walk the *levadas*: some routes are scarily vertiginous, and include long tunnels.

The point at the far eastern end of Madeira, the Ponta de São Lourenço, is one of the most arid parts of the island, but comes alive in spring with a carpet of buttercups. The point was named after the ship of the 15th-century explorer João Gonçalves Zarco.

Grapes that are used to make Madeira's famous fortified wines are grown on tiny terraces called *poios*, many of which were cut out of the hillsides by early settlers. There are four main types of Madeira wine: Malmsey (or Malvasia), Bual, Sercial and Verdelho. Despite the deep colour of the wine, most of the grapes used are white.

and other fruits more readily associated with tropical climates grow at sea level, while strawberries and grapes grow at higher elevations. Madeira is famous for its flowers, which range from bougainvillea and jacarandas to forget-me-nots and heathers.

Grapes grown on hillside terraces are used to make Madeira wine. Its rich, nutty flavour, so the story goes, dates back to about 1700 when a ship returned to Madeira from a trip through the heat of the Equator with an unused cask of wine. To the surprise of all, the wine was deemed to have been improved by this experience. A method of replicating it was evolved by gently heating wine for six months – a process called *estufagem*. The wine was fortified with brandy to prevent it from spoiling. The British were very fond of Madeira wine, and became closely involved in the trade, notably after Britain took control of Madeira in 1807–14, during the Napoleonic Wars.

Today tourism is Madeira's most important source of revenue, with a million visitors coming each year. But, because it has almost no beaches to speak of, Madeira has never been a mass-market destination. Its charms lie in its constant and agreeable weather, beautiful landscape and dignified traditions, such as its embroidery and wickerwork, its saints' day festivals and quaint oddities such as the *carros de cesto* – wicker sledges that run down the road from Monte into Funchal.

The Ancient Greeks had some notion of the existence of islands beyond the Strait of Gibraltar. They called them Macaronesia, meaning 'Isles of the Fortunate', a strangely prescient label that few visitors today would argue with.

BORNHOLM

Latitude 55°5'N **Longitude** 14°43'

Area 588 square kilometres (227 sq miles)

Status Island of Denmark

Population 42,200

Main settlement Rønne

Official language Danish

Currency Danish krone

Bornholm is known as the 'Pearl of the Baltic', a reference to the bright, opalescent light that bounces off its white-sand beaches and the surrounding sea, and the beauty of its gentle seaside towns. It also boasts more hours of sunshine that any other place in Denmark.

Although the island is Danish, the nearest part of Denmark lies 140 kilometres (87 miles) to the west, and Bornholm is actually closer to Sweden, Germany and Poland. It therefore attracts a very international crowd. They come in summer, by

TRAVELLER'S **TIPS**

Best time to go: Summer is best, especially June–August. Winter may be icy, and many of the main attractions close November–March.

Look out for: The Medieval Centre of Bornholm holds a big Medieval Market in late July, with jousting, feasts, entertainment and battle reenactments.

Dos and don'ts: Do go cycling. There are 200 kilometres (125 miles) of cycling trails on the island, and plenty of facilities to hire bikes.

The beach at Dueodde, at the very southern tip of this lozenge-shaped island, has remarkably soft, white sand. It might be a tropical island, but for marram grass growing in the dunes.

flights and by ferries, to camp, sail, windsurf, walk, cycle, go horseriding, swim or just lie on the beach and sunbathe. Seafood is another great attraction, with smoked herring at the top of the list: there are smokehouses in many seaside towns.

Bornholm is geared up for its summer influx of family holidaymakers. The Joboland amusement park at Svaneke offers rides, a boating lake and a children's zoo. The Sommerfuglpark at Nexø is a large, indoor tropical butterfly farm. Near Østerlars is the Bornholms Middelaldercenter (Medieval Centre of Bornholm), a reconstructed medieval village, animated by craftsmen and performers.

Since the Viking era, Bornholm has been coveted by the Swedish and the Germans, and it has changed hands many times. This explains the robust structure of the island's four unusual round churches, all built between 1150 and 1200, now kept pristinely whitewashed. It also explains the number of castles. The ruined Hammershus, in the very north of the island, is the largest medieval fortification in northern Europe.

KEFALONIA

Latitude 38°15'N **Longitude** 20°30'E

Area 781 square kilometres (302 sq miles)

Status An Ionian island of Greece

Population 36,400

Capital Argostoli

Official language Greek

Currency Euro

The Ionian Islands lie to the west of mainland Greece, stretching down from Corfu in the north to Zakynthos in the south. Between them, snuggled up close to Ithaca, is the largest of them all: Kefalonia.

This sprawling island is shaped like a swimming frog, its back legs formed by two peninsulas. One stretches out towards the northern port of Fiskardo, the other folds back around the island's second town of Lixouri, on the Gulf of Argostoli. Deep bays on every side offer views along the coast, across an aquamarine

sea, to the dry hills softened by the smoky distant light. These hills rise up to the island's highest point at Mount Ainos, at 1,628 metres (5,341 ft).

Although much of the landscape consists of arid limestone, there is grazing for goats and swathes of woodland that include Kefalonia's own fir tree, the *Abies cephalonica.* There are vineyards where grapes are grown to make wine and currants, and scores of olive groves used to produce olive oil, one of the island's chief exports.

Kefalonia was an unsung tourist destination until the recent past. The 20th century had been brutal. Occupying Italian and German troops clashed in 1943, resulting in the cold-blooded massacre of some 5,000 Italian soldiers. Then in August 1953 a calamitous earthquake destroyed the majority of buildings on the island; only Fiskardo on the northern peninsula survived unscathed.

But in 1994 the British author Louis de Bernières published his haunting novel *Captain Corelli's Mandolin,* a love story set

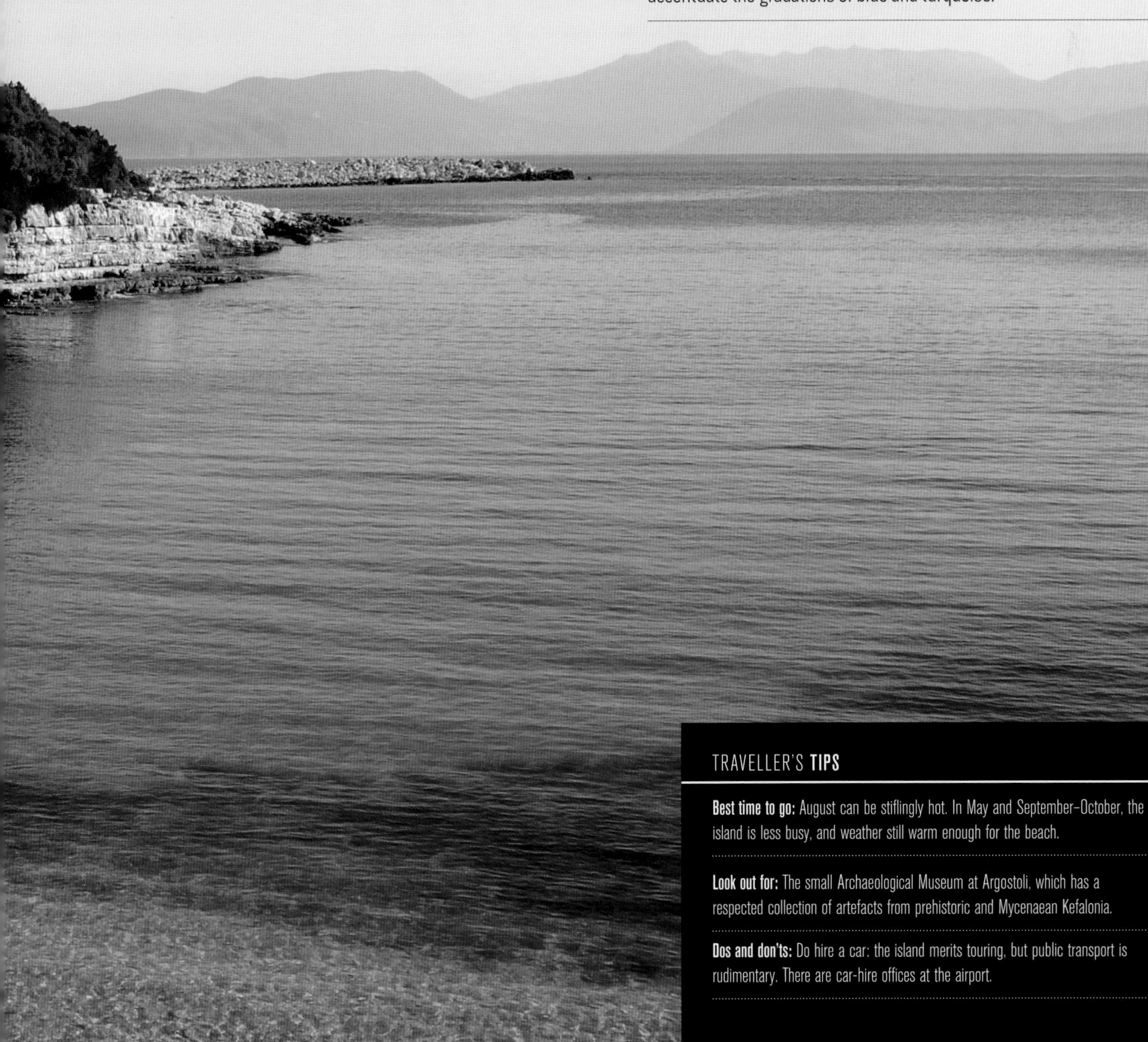

Myrtos Bay, on the northwest coast, has regularly been cited as one of the most beautiful beaches in the world. Pebbles of white marble accentuate the gradations of blue and turquoise.

TRAVELLER'S **TIPS**

Best time to go: August can be stiflingly hot. In May and September-October, the island is less busy, and weather still warm enough for the beach.

Look out for: The small Archaeological Museum at Argostoli, which has a respected collection of artefacts from prehistoric and Mycenaean Kefalonia.

Dos and don'ts: Do hire a car: the island merits touring, but public transport is rudimentary. There are car-hire offices at the airport.

in Kefalonia during those grim days of the Second World War, but nonetheless painting an affectionate and appealing portrait of the island. A runaway bestseller, largely through word-of-mouth, the novel was made into a movie starring Nicholas Cage and Penélope Cruz in 2001 – and filmed in Kefalonia. The outstanding beauty of the island landscape reached an international audience, and travellers came from all over the world to see it for themselves. The island proved big enough to handle the increased numbers: today it still seems relatively empty and unpressured.

The beaches and landscape remain the primary attractions. There are stalactite caves at Drogarati; and at Lake Melissani, near Sami, rowing boats take visitors into a cave where the water turns an astonishing blue. Many of the southern beaches on Kefalonia are rare nesting sites where endangered loggerhead turtles lay eggs between June and early August – an activity that is now carefully monitored and protected.

For those who want to delve into Kefalonia's very long history there are archaeological sites that reach back through 40,000 years. Among them are Mycenaean tholos tombs (built like round stone shelters) dating back to 1300 BC. A major recent find of Roman graves near Fiskardo has shed new light on the era of their occupation, which lasted from the second century BC to the fourth century AD.

Since that time, Kefalonia has been invaded and occupied without cease. The Normans of Sicily were dislodged by the Norman adventurer Robert Guiscard, who died on the island

in 1085 – and after whom the port of Fiskardo was renamed. Then came the Pisans, and then the Venetians, who held the island from the 13th century to 1797 (with only one brief interruption by the Ottoman Turks in 1479–1500), and left behind several castles, including the Castle of St George at Kastro. The Venetians were ousted by the French, who were removed in turn by the British in 1809. The British stayed on the island until 1864, building roads and bridges, as well as the original version of the famous neoclassical Agios Theodori Fanari (lighthouse) near Argostoli.

But the 1953 earthquake has left the darkest shadow, not only because it tore down most of the island's historic buildings but because it caused a large proportion of the population to emigrate. It would be tempting to suggest that Kefalonia can rebuild, riding on the success of its newfound popularity. Unfortunately it sits on top of a tectonic faultline, and fresh, if milder earthquakes are a recurring event. This, however, cannot detract from the unquestioned beauty of the landscape.

A fishing boat and yachts lie at rest next to the quays at the pretty little port of Fiskardo, on the northernmost tip of Kefalonia. Ferries leave from here to cross the sound to Ithaca.

SANTORINI

Latitude 36°25'N **Longitude** 25°26'E

Area 91 square kilometres (35 sq miles)

Status Island of the Cyclades Group, Greece

Population 12,500

Main settlement Thira

Official language Greek

Currency Euro

Santorini is unlike anywhere else on earth. From the deck of an incoming ferry, the sky suddenly fills with the towering black walls of a volcanic caldera, embracing a scoop of sea of an astonishingly deep blue. The whitewashed houses of main town, Thira (also a name used for Santorini as a whole) are piled up along the volcanic cliffs, high above the sea, like icing on a chocolate cake.

It made sense to build up here when the ships of pirates and competing empires plied these waters – which was throughout most of history. The Cyclades islands – created, according to legend, when the sea god Poseidon struck his trident against the rocks of mainland Greece – are in the middle of trade routes that connected the Mediterranean world. This is why the Minoans settled here, as an outpost of their base in Crete. At Akrotiri on Santorini, excavations have revealed an extensive Bronze Age Minoan city, complete with wall-paintings depicting vividly its life and times. But the city was abandoned quickly, it seems, some time around 1500 BC, and then covered in a thick layer of volcanic ash. For, at about that time, a colossal volcano erupted, blasting out the huge, collapsed crater that is Santorini today.

Santorini offers a similar kind of holiday experience to any Greek island, with excellent beaches, water sports and friendly tavernas serving fresh grilled fish and intense local wines. But it is impossible to ignore the drama of the landscape, or the big question: was this the site of Atlantis, the fabled lost civilization – which, as Plato put it, 'in a single day and night of misfortune… disappeared in the depths of the sea'?

Blue church domes contribute to Oia's renown as perhaps the most beautiful town on Santorini, famed also as the place to watch what are claimed to be the world's loveliest sunsets.

TRAVELLER'S **TIPS**

Best time to go: The season lasts from March to November. March is recommended for its lack of heat and crowds. Many businesses close December–February.

Look out for: Day trips (or overnight stays) on the island of Thirassia, in the arc of Santorini. Unspoilt and quiet, it is a vision of how the Greek islands used to be.

Dos and don'ts: Walk up the 588 zigzagging steps to Thira at your peril: the donkeys carrying goods may barge you. The cable car is a far easier option.

CRETE

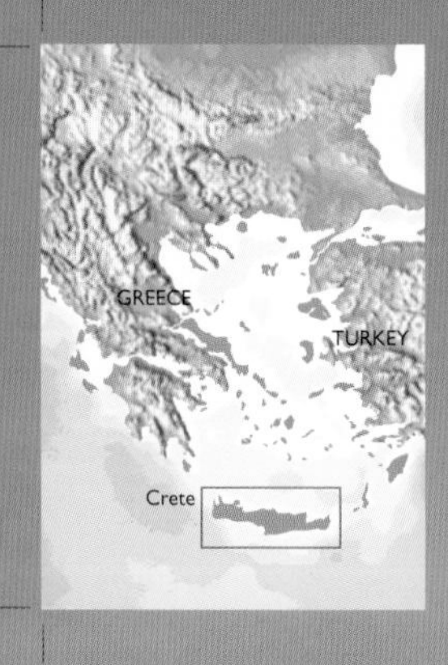

Latitude 35°12'N **Longitude** 24°54'E

Area 8,336 square kilometres (3,219 sq miles)

Status Island of Greece

Population 624,000

Capital Heraklion

Official language Greek

Currency Euro

Greek myth becomes a physical reality in Crete. The braying goats, the cicada-like tzi-tzi singing of the tzitzikia flies from the trees in the hot summer months, the plaintive cry of a lone hawk – all would all have been familiar to the Ancient Greek hero Theseus as he rode the cobalt seas into some sheltered cove and headed inland to the palace of King Minos.

There is a timeless beauty to the rugged, green and mountainous landscape, which naturally suggests sanctity. Zeus, father of the gods, is said to have been born in a cave

TRAVELLER'S **TIPS**

Best time to go: Summer lasts from April to November, and can be very hot in midsummer. In spring (April) the landscape is awash with wild flowers.

Look out for: Cretan music, played on a *lýra* (three-stringed fiddle held vertically) and *láoúto* (lute), with singing that uses quarter-tones in Middle-Eastern fashion.

Dos and don'ts: Do go to the Heraklion Archaeological Museum, where many of the Minoan treasures are stored, including evidence of the sport of bull-jumping.

on Mount Ida, Crete's highest peak. Now called Mount Psiloritis, it rises to 2,456 metres (8,058 ft) – high enough to ensure that winter snow stays on its peak until June. Zeus disguised himself as a bull to abduct Europa, and she gave birth to Minos. As a punishment for failing to make a bull sacrifice to the sea god Poseidon, King Minos had to look after the terrible monster – half man, half bull – called the Minotaur, which he kept in a labyrinth beneath his palace at Knossos, feeding it with youths exacted as tribute from Athens – until Theseus killed it, and then found his way out of the

The Palace of Minos at Knossos is now a major visitor attraction, and includes some sections of reconstruction, such as these red columns and murals, which give an indication of the bright colours of the original buildings. The complex includes a throne room, banqueting hall, shrines and extensive storerooms.

Chania, in the northwest of the island, was an old Venetian port. The harbour today is overlooked by tavernas (traditional restaurants), while behind lies a web of tiny streets lined with houses from Crete's Venetian and Turkish past.

The 'Iron Gates' mark the narrowest point of the Samariá Gorge, where the River Tarraíos passes though a gap just 3 metres (10 ft) wide at the base of cliffs rising to some 500 metres (1,600 ft). The massive boulders are an indication of the power and turbulence of the river when in full flood in early spring.

labyrinth again with the help of a thread given to him by King Minos's daughter Ariadne, who had fallen in love with him.

In 1900 the British archaeologist Arthur Evans began excavations at the site of Knossos, near Heraklion, gradually unearthing a huge palace complex. This was the centre of the first major European Bronze Age civilization, which lasted from about 2000 to 1350 BC; Evans called it Minoan, after King Minos. He may not have found the labyrinth, but there was plenty of evidence of a bull cult at Knossos.

After the fall of the Minoans, Crete was ruled by all the empires that have held sway over this region: Mycenaeans, Greeks, Romans, Byzantines, Venetians. The Ottoman Empire took the island from the Venetians in 1669 and ruled it to 1897, during which time a large percentage of the population became Muslim. The legacy of these eras is written in the churches, castles, minarets and town architecture of the island today.

But Crete is also an island where nature holds human activity firmly in balance, and there is no more dramatic example of this than at the Samariá Gorge, now protected as a National Park. Here a river has cut a deep path through the mountains, creating an 18-kilometre (11-mile) walk that squeezes through narrow pairs of cliffs soaring over 500 metres (1,600 ft). The path is open only from May to October: in winter the river is a gushing torrent filled with meltwater from the Lefka Ori Mountains.

As the Cretan author Nikos Kazantzakis suggested in his novel *Zorba the Greek* (1946), set – and later filmed – in Crete, his fellow islanders are proud of the land and history that have shaped them in such a distinct fashion, and they cling to the many aspects of the rich culture that sets them apart: their dances, music, traditional costumes, the Cretan Greek dialect and poetry. Their cheeses, made from goat's and ewe's milk, are also distinctive, such as the hard and nutty *graviera*, or the soft, fresh cheese called variously *anthotiros* and *mizithra* – flavours that Theseus, too, would have recognized.

Church bells call the faithful to prayer in the valley of Festos (or Phaistos), close to where St Paul came on his journey to Rome in about AD 59. St Titus, first bishop of Crete, was one of St Paul's companions.

SURTSEY

Latitude 63°18'N **Longitude** 20°36'W

Area 1.4 square kilometres (0.54 sq miles)

Status Island of Iceland

Population 0

Main settlement None

Official language Icelandic

Currency None

On 15 November 1963 a cook on board a fishing trawler, passing the Westman Isles (Vestmannaeyjar) off the southwest coast of Iceland, noticed a plume of smoke rising out of the sea. He raised the alarm, and the trawler headed over to investigate, expecting to find a ship on fire. Instead they discovered nothing but sulphurous, black fumes belching forth out of the waves.

Within a few weeks a volcano, rising from the seabed 130 metres (426 ft) below, had pushed its head above the water, blowing a column of smoke 10 kilometres (6 miles) into the sky. Explosions, caused by seawater falling into the vents, hurled large rocks hundreds of metres out to sea, while a steady emission of lava and volcanic debris called tephra began to form an island, battling against the swell and surge of the waves. This brand new island was named Surtsey, after the fearsome giant of Icelandic myth, Surtr, who struck with his flaming sword and laid waste to everything in his path.

As Surtsey continued erupting, two other small islands appeared, but they fell back beneath the waves as soon as activity ceased. Surtsey, however, was made of harder stuff. When finally, on 5 June 1967, its eruptions ceased it was an island measuring of 2.7 square kilometres (1 sq mile), and rising to 174 metres (570 ft). Since then – through the action of the waves and a natural sinking effect as the volcano compacts under its own weight – it has reduced to about half that surface area and lost a little height, but the core of what remains today will survive.

The fascination of Surtsey now lies in the manner in which it becomes colonized by plants and animals. Declared a Nature

Waves crash onto the shores of Surtsey, where access by human beings is restricted to a few visiting scientists.

Atlantic puffins are common in the Westman Isles, and some were found nesting on Surtsey in 2004. They usually nest in burrows excavated from grassy cliffs, but may nest in scrapes among rocks where gulls and other predators are less of a threat. This may explain why they did not colonize Surtsey until 37 years after eruptions ceased.

Reserve in 1965, Surtsey has provided a rare blank canvas on which to watch the process called biocolonization. With mainland Iceland just 35 kilometres (22 miles) away, and islands in the Westman archipelago even nearer, biocolonization has been rapid. First came the birds: gulls landed on the island while the volcano was still erupting. By 1970, fulmars and guillemots were nesting; puffin nests were found in 2004. Migratory birds – such as whooper swans and geese – regularly drop by.

Birds bring seeds and insects on their bodies and in nesting material, and their guano fertilizes soil. So next came plant life, beginning with mosses and lichen, followed by ferns, grasses and shrubs. More than 30 species of plant have now achieved a permanent foothold. Insects and spiders have been carried on the wind; and various forms of life – slugs, beetles, worms – are cast up on Surtsey's shores clinging to driftwood or lumps of turf, or on animal carcasses. Limpets, sea-urchins, starfish, crabs and seaweed have colonized the shore. Harbour seals and grey seals began breeding here in the 1980s.

The only permanent human intrusion is a hut used by visiting scientists – vulcanologists, zoologists, botanists. Surtsey has provided a unique laboratory that shows the irresistible power of life to find a niche in what – on the face of it – looks like one of the most hostile environments on Earth.

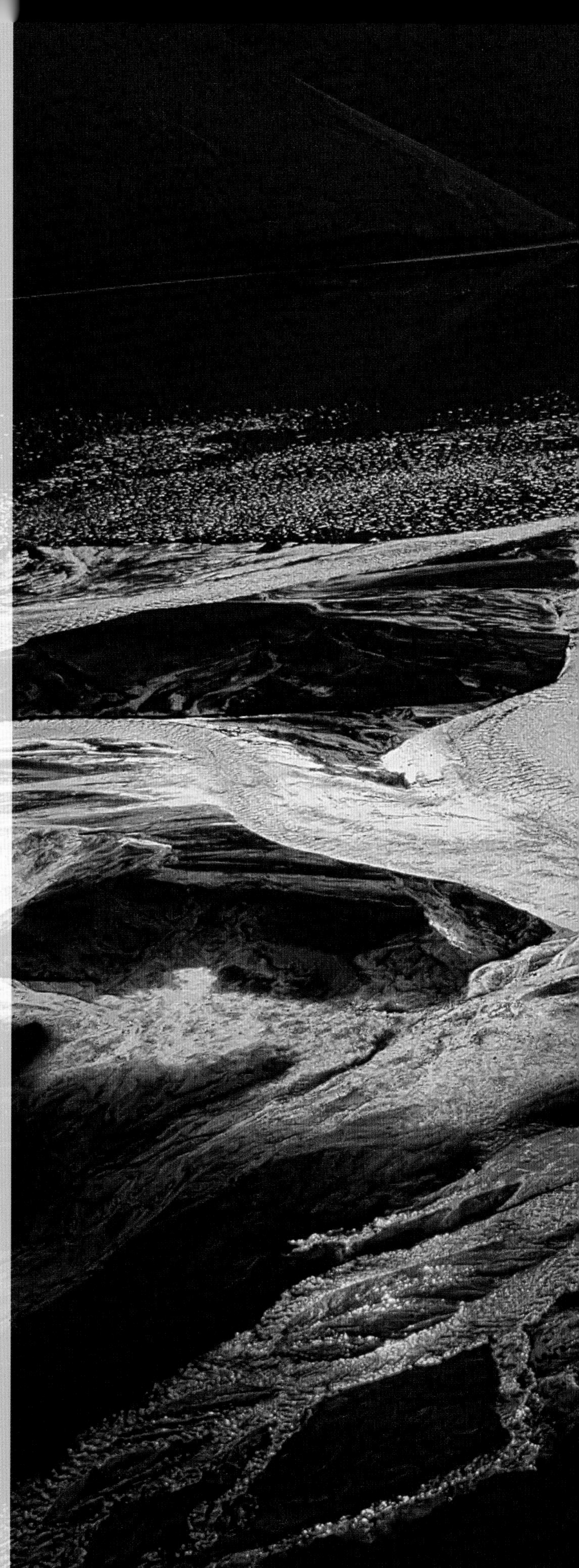
The sterile terrain of Surtsey creates strange forms all of its own. Here, at the edge of the volcanic crater, the laval rock resembles a flood of molten caramel or a river of chocolate that has burst its banks.

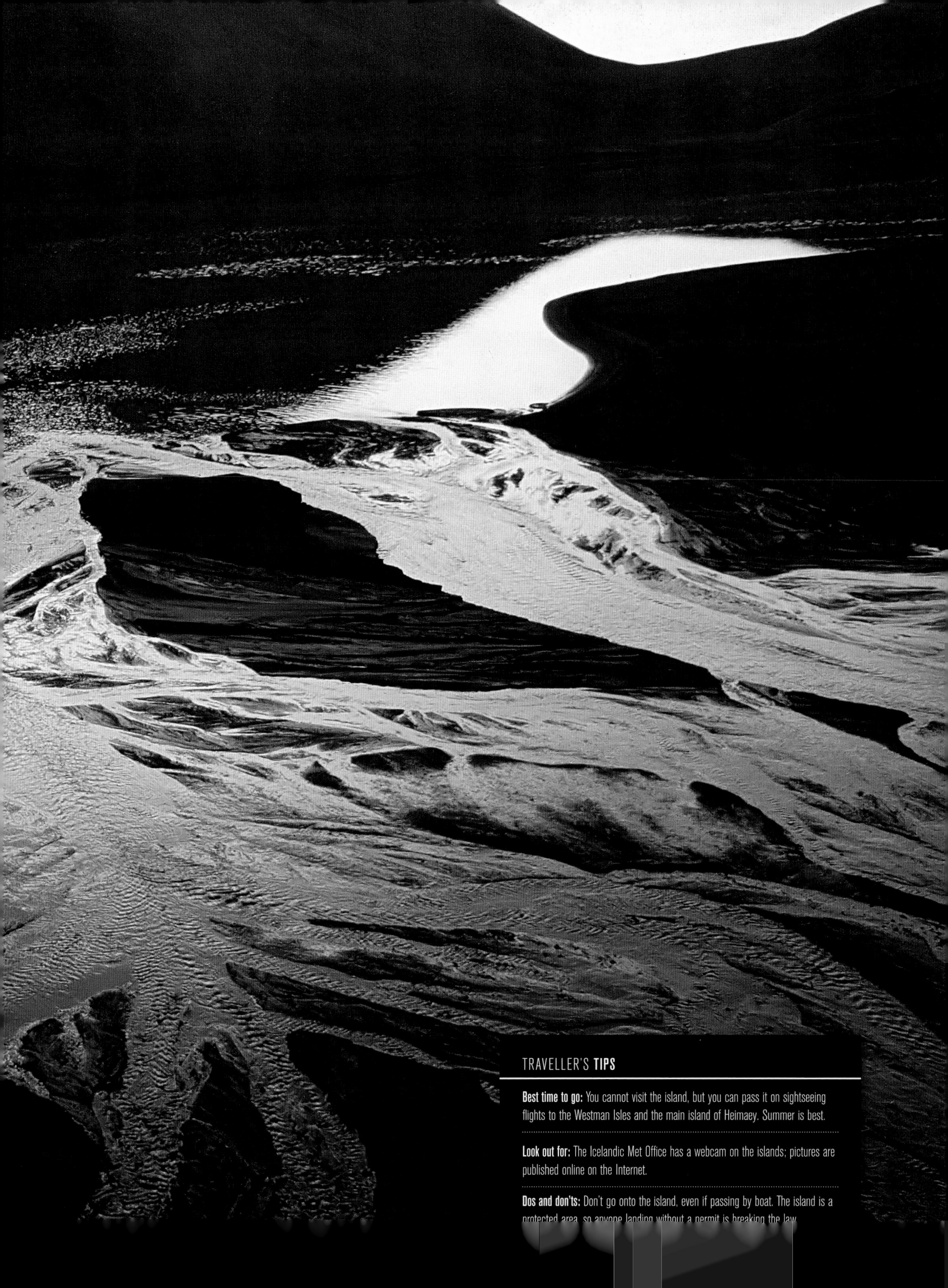

TRAVELLER'S **TIPS**

Best time to go: You cannot visit the island, but you can pass it on sightseeing flights to the Westman Isles and the main island of Heimaey. Summer is best.

Look out for: The Icelandic Met Office has a webcam on the islands; pictures are published online on the Internet.

Dos and don'ts: Don't go onto the island, even if passing by boat. The island is a protected area, so anyone landing without a permit is breaking the law.

LOFOTEN ISLANDS

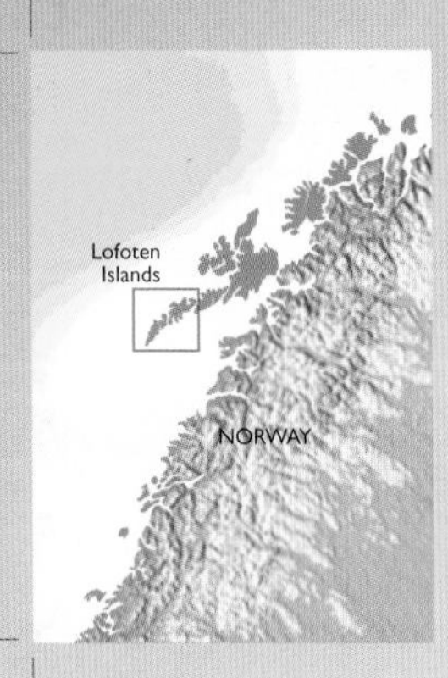

Latitude 68°14'N **Longitude** 14°34'E

Area 1,227 square kilometres (474 sq miles)

Status Archipelago in the county of Nordland, Norway

Population 24,500

Main town Svolvær

Official language Norwegian

Currency Norwegian krone

The Lofoten Islands lie within the Arctic Circle, on the same latitude as the ice-bound northern coasts of Canada and Alaska. Yet, here, warmed by the Gulf Stream, temperatures rarely dip much below freezing – even in a land where the sun does not appear above the horizon at all for a month in midwinter.

The islands stretch like pearls on a string over a distance of some 240 kilometres (150 miles) from the western tip of Hinnøya, through the larger islands of Austvågøy and Vestvågøy to the smaller Flakstadøy and Moskenesøya. Then, further out to sea,

lie Værøy and Røst, and a sprinkling of smaller islands. A road, the E10, now runs right through the islands from mainland Norway, linking them with bridges and tunnels until it reaches Moskenesøya, and the place called, simply Å (it means 'stream').

Formed of a mix of ancient and newer rock, sculpted by the ice ages, the islands are a set of jagged peaks, rising to nearly 1,200 metres (3,940 ft), which, when approached from the sea, look like a single, dark and massive wall. Only up close can they be seen to be fissured by fjords and narrow straits.

TRAVELLER'S **TIPS**

Best time to go: The best weather lasts from May to September. October is the wettest month. In December to early January there is no sunlight at all.

Look out for: The Lofotr Viking Museum at Borg, on Vestvågøy. Viking life and historical artefacts are displayed in a reconstruction of a huge Viking longhouse.

Dos and don'ts: Do bring clothes to match very uncertain weather. This is the Arctic Circle: it can be cold and wet at all times of the year.

Sakrisøy, on the island of Moskenesøya, is reflected in mirror-still conditions of the Vestfjord. Some of the old fishing buildings have now been converted into hotel accommodation.

Off the coast of the Lofoten Islands, a killer whale, or orca, lifts its head out of the water to 'spy-hop' – a habit a little like treading water, which allows it to look for seals to prey on, or to observe boats. Whale watching is one of the great attractions of the Lofoten Islands, and minke, humpback, pilot and sperm whales may also be seen.

The name Lofoten is what Vestvågøy alone used to be called, because it apparently resembled a lynx (*lô*) foot (*fótr*). Beneath the towering peaks, hugging the rocky shores and sandy coves, lie settlements of painted rectangular homes, warehouses, churches – scattered like toy building bricks. Many of these are modelled on traditional cabins on stilts called *rorbuer*, used as seasonal lodgings by fishermen: *bu* for small dwelling, and *ro* a reference to the rowing boats they always used until the introduction of motor engines. The settlements tend to be located on the 'inner' (south and southeastern) sides of the islands, rather than the 'outer' sides that face towards the Norwegian Sea, because that is the source not only of violent storms but also a thick, cold 'good-weather fog' that can blanket the western shores in summer.

Typical red *rorbuer* cabins perch on their stilts above the rocky shores of Reine, on Moskenesøya. In the season (which lasts from February to early June), Reine is a good place to see cod hung out to dry on wooden racks, to make the famous air-dried stockfish of the Lofoten Islands.

Most settlements have harbours and quays lined with boats. Fishing is, and always has been, a major activity of these islands, for, just to the south of the Lofoten Islands lies the Vestfjord, a major spawning ground for cod. January to April is the fishing season, and the cod are caught by net and line, and then air-dried on racks in a traditional manner perfected during the Viking era more than 1,000 years ago. Cod 'stockfish' preserved in this way can be kept for years. Stockfish is used locally to make a variety of dishes such as the pungent *lutefisk*, traditionally associated with Christmas. For centuries, it was also exported, through Bergen and the Hanseatic League, right across Europe. So, too, was cod liver oil.

The Lofoten Islands, far from being isolated, had connections to the world, and today they are a popular tourist destination. The islands are famed for their beauty, attracting hikers and cyclists as well as artists and sculptors. Summer temperatures can be warm enough to encourage visitors to head for the beach. The alpine landscape is cherished by climbers, who can be out at all hours of the day and night in summer.

In Svolvær, the sun does not set at all between 25 May and 17 July. Many visitors come specially to see the midnight sun; others come in winter in the hope of seeing the Northern Lights, or Aurora Borealis, which can fill the sky with its eerie play of colours at any time between September and April.

SPITSBERGEN

Latitude 78°54'N **Longitude** 18°1'E

Area 39,044 square kilometres (15,075 sq miles)

Status Island of Norway

Population 2,500

Capital Longyearbyen

Official language Norwegian

Currency Norwegian krone

'The Wild West of the Arctic' is how Spitsbergen has been described. Whalers, fur-hunters, coal miners, polar explorers, and now oil prospectors, have tried their luck here, battling against extreme conditions, polar bears and each other. They have left their marks, but these are small scars in a huge land ruled by the overwhelming forces of nature.

Spitsbergen is the largest island in the Svalbard archipelago. It lies in the Barents Sea 650 kilometres (400 miles) to the north of Norway, and about 1,300 kilometres (800 miles) from the

TRAVELLER'S **TIPS**

Best time to go: The main season is June–August, which coincides with the midnight sun. The winter sports season in March–May.

Look out for: The Svalbard Museum in Longyearbyen covers the history of the islands, including many of its colourful episodes of derring-do, and disasters.

Dos and don'ts: Don't ever approach polar bears. They may look cumbersome and cuddly, but you are food and they can run at 40 kilometres per hour (25 mph).

In summer the snows melt, revealing a khaki tundra. The Dutch explorer Willem Barentsz gave Spitsbergen its name in 1596: it means 'pointed mountains'.

North Pole. This would be an icy wasteland were it not for the warm Gulf Stream and its extension, the North Atlantic Current, which moderates the extreme temperatures normally found in these latitudes. As it is, summer temperatures Spitsbergen average 5°C (41°F), and winter -14°C (7°F). Polar winter brings utter darkness for nearly four months of the year, from 26 October to 15 February; but from 20 April to 23 August there is light all the time.

This midnight sun provides sufficient light for a remarkable range of plants – some 165 species in all – such as the delicate

Polar bears roam the island, and are one of the great attractions of Spitsbergen. Given that they can weigh up to 500 kilogrammes (1,100 lbs), and do not count humans as friends, they are treated with respect. Anyone travelling outside the settlements is required to be armed with a rifle, and is permitted to shoot in self-defence.

purple saxifrage and Svalbard poppy. They populate the soggy tundra, helping to attract 30 species of migratory birds. Many of these come to nest here, in their millions. The most common are auks, fulmars, kittiwakes, guillemots and ivory gulls. Only one bird overwinters: the rock ptarmigan. The sea around Spitsbergen has seals, walruses and beluga whales. On land there are Arctic foxes, reindeer and polar bears.

Norway has owned Spitsbergen since 1905, a status confirmed by the Svalbard Treaty of 1920, but the 40 signatory countries have a right to exploit the islands commercially. It was not always thus. Since Willem Barentsz first stumbled upon Spitsbergen in 1596 a succession of countries have tried to stake a claim. Barentsz was attempting to find a Northeast Passage – a way through the ice to reach the trading ports of the Far East. Others followed, including the Englishman Henry Hudson in 1607.

So the Dutch first claimed the islands, then the English, then the Danes. Beluga whales, seals and polar bears were the initial lures, for their valuable blubber and furs. Whaling and hunting carried on into the 20th century; there are bays in Spitsbergen still littered with bleached whale bones from the slaughter. But during the 19th century coal was found and mined. Squabbling among the competing coal-mining companies led eventually to the call for the international settlement realized in the Svalbard Treaty. During the latter half of 20th century the turmoil calmed.

Settlement is now concentrated in just four 'towns', linked by sea, air – or snowmobile outside summer (there are no connecting roads). Russia runs a coal mine at the town called Barentsburg – a Soviet-era relic noted for its propaganda murals, and the world's most northerly statue of Lenin. The mining may no longer be viable, but it is widely thought that

Russia is determined to maintain a presence, because – according to some estimates – the Svalbard archipelago stands above one quarter of the world's oil reserves, as yet untapped.

Tourism is a major activity now. Some pursue wintersports in spring, arriving by air. Many summer visitors are on cruise ships, and tour the fjords and glaciers to admire the astonishing beauty of the landscape and its wildlife – fog and icebergs permitting. And they will also stop at some of the huts left abandoned in impossibly remote places – the forlorn remnants of human endeavour in the Arctic's Wild West.

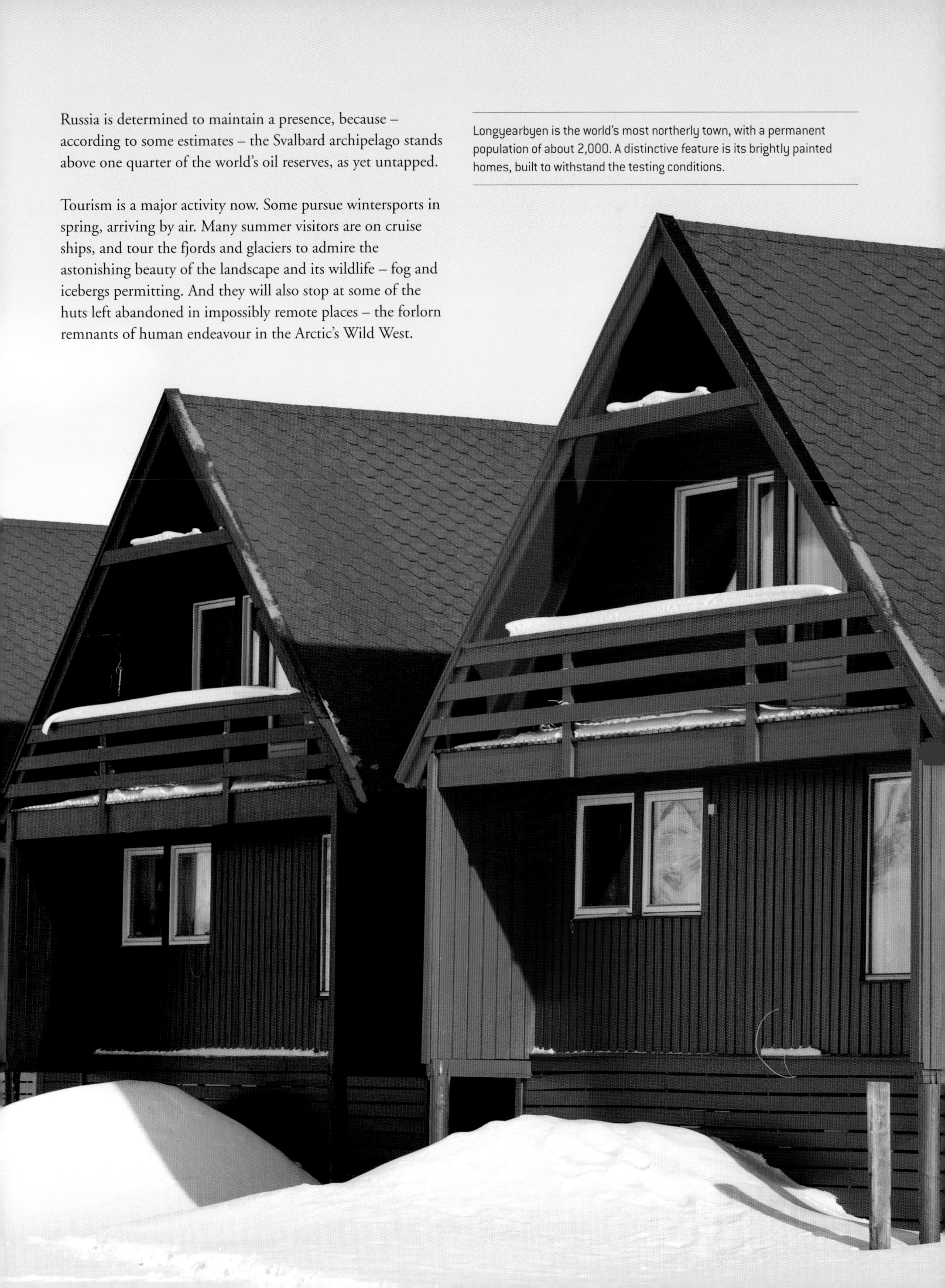

Longyearbyen is the world's most northerly town, with a permanent population of about 2,000. A distinctive feature is its brightly painted homes, built to withstand the testing conditions.

HVAR

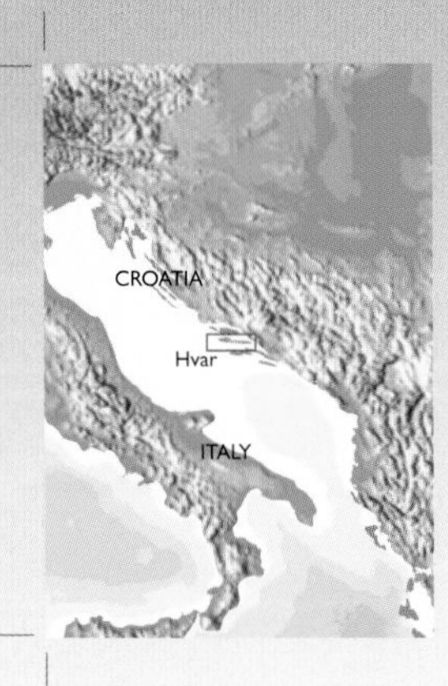

Latitude 43°8'N **Longitude** 16°44'E

Area 297 square kilometres (115 sq miles)

Status Island of Croatia

Population 11,100

Main settlement Hvar

Official language Croatian

Currency Kuna

'If you know Hvar, you know heaven,' goes a local saying. And there are many aspects to this heaven. Hvar is a hugely popular holiday destination, with a youthful club scene during the summer months. But it is also the 'Riviera of the Dalmatian Coast', with excellent hotels and chic bars, and a high-summer season that attracts international filmstars and sportstars, and fills the beautiful harbour of Hvar City with luxury yachts. The climate is also a great part of the appeal. The number 2,718 is widely quoted: that is the hours of sunshine per year – more sunshine, apparently, than any other place in the Mediterranean.

Hvar is dubbed the 'Island of Lavender' because of its lavender-growing industry; it developed in the 19th century, feeding into the French perfume business. Hvar lavender oil and soap are popular souvenirs.

TRAVELLER'S **TIPS**

Best time to go: High season lasts from May to September; in July–August Hvar City is heaving, but fun. September–October is quieter, and can still be warm.

Look out for: Hvar is one of the two regions in Croatia noted for wine. Its excellent reds and whites rarely leave its shores: you have to visit to drink them.

Dos and don'ts: Don't panic if you feel wretched when the jugo wind (the sirocco) blows in from the Sahara, usually in spring and autumn: everyone suffers.

To reach Hvar you have to come by boat – by yacht or by ferry. That reinforces Hvar's identity as an island, and the sense of leaving behind another world to reach it. Ferries depart from the bustling city of Split to the north, an hour's journey, skirting around the island of Brač.

Views of Hvar from an approaching ferry hint at its special beauty: a long line of rolling limestone hills dappled with green. A cigar-shaped strip of land, sticking out horizontally into the Adriatic Sea from close to the mainland, Hvar has an unusually fertile landscape for these islands, blessed by plentiful springs and streams. The interior has pine forests, vineyards, olive groves and fruit orchards.

This is what has attracted people to the island since the Stone Age. Greek settlers came here and, in 384 BC, founded a port on the north coast that is now Stari Grad: where the car ferries dock today is actually one of Europe's oldest towns. The Ancient Greeks also laid out a field system on the Stari Grad Plain, divided by walls and with cisterns for collecting rain water: it is still in use today, 24 centuries later. But it was the Venetians who left the greatest mark on Hvar, particularly at

their main port, Hvar City, where the passenger ferries arrive. Set around a beautiful harbour, and its vast central piazza, Trg Svetog Stjepana (St Stephen's Square), is a medieval city elegantly built and paved in white limestone – it was once a naval base for Venice's fleet of galleys.

The strategic importance of the island is underlined by the 16th-century fortress that hovers above Hvar City, known as the Spanjola after its Spanish architects; this was part of the Venetian defences against pirates, and the Ottoman Turks. The serene 15th-century Franciscan Monastery in Hvar City, built as a retreat for sailors, reveals a more contemplative side to island life: a 500-year-old cypress tree growing in a courtyard serves as a symbol of how history lives on here.

When the Napoleonic Wars put an end to the Venetian empire, Hvar became part of the Austrian empire, and fell into decline. But this also preserved it from development. Its new prosperity has funded careful and sensitive restoration: many of the boutiques, hotels, restaurants and cafés are housed in buildings 400 years old. Now the shores buzz with motorboats, yachts, dinghies and windsurfers. The beaches on

the island itself are limited, but just outside the entrance to Hvar City's harbour are the Pakleni Otoci, a string of 16 little islands where there are beaches, and beach restaurants serving lunch of freshly grilled seafood – particularly squid. So summer visitors hire boats – water taxis or self-drive – and head out to the islands for the day, finding a quiet spot in one of the many coves and inlets, among the pines, to relax, swim and snorkel, and absorb some of those 2,718 hours of sunshine. Then they motor back across the bay, and dress up to join the chic crowd in the floodlit town, and to revel in the vibrant nightlife for which Hvar is now famous.

The view from beneath the ramparts of the Spanjola fortress, at the northern edge of Hvar City, shows the typical red tiles of the rooftops, and the open, 16th-century belfry of the Church of St Mark. In the distance lies Galešnik, one of the smallest and easternmost of the Pakleni Islands, and the closest to Hvar's harbour.

Bars and restaurants spill out onto Trg Svetog Stjepana, a grand public space almost the size of a football pitch, beside the harbour of Hvar City. To the right is the 16th-century Arsenal, and in the distance the 16th–17th century façade of the Cathedral of St Stephen.

KORNATI ISLANDS

Latitude 43°47'N **Longitude** 15°20'E

Area 320 square kilometres (126 sq miles)

Status Islands of Croatia

Population No permanent settlements

Main settlement None

Official language Croatian

Currency Kuna

There is a special, hard-won beauty to the dry limestone islands of the Adriatic. Azure seas salt-lick the sunbaked rock. Dry grasses, smelling of hay, cling to scarce footholds among acres of exposed karst stone that sucks up any rain into its brittle, fissured surfaces. Grasshoppers chirrup and skip, lizards skitter and butterflies flicker in search of tiny wildflowers on woody stems – life forms able to prosper on dew.

The Kornati Islands lie like a long streak of thin paint about 10 kilometres (6 miles) off the Dalmatian coast of Croatia.

There are 140 of them, forming the most densely packed group of islands in the Mediterranean. The largest is Kornat, 25 kilometres (16 miles) long, and no more than 2.5 kilometres (1½ miles) wide.

None of the islands is permanently inhabited, but there are clusters of dwellings here and there, nestling in the many coves and inlets, that are used as seasonal accommodation by fishermen and farmers from neighbouring islands – or, these days, by holidaymakers.

TRAVELLER'S **TIPS**

Best time to go: Summer (July–August) is the high season, when temperatures average 25°C (77°F). There are clear skies 151 days of the year; most rain falls in October.

Look out for: Marked paths lead up to the peaks (referred to as 'belvederes'), which offer unparalleled views over the island chain; for example, at Opat, south Kornat.

Dos and don'ts: You must have a permit to enter the National Park. Permits can be purchased at the Park offices at Murter (a neighbouring island), or at marinas.

Some 75 sailing boats with traditional lanteen sails gather for a regatta off the Kornati Islands. For centuries, these were the typical boats of the islands, used for transport and fishing.

The limestone was laid down by marine sediment millions of years ago, but tectonic uplift has created cliffs, known locally as 'crowns', on the southern ends of some of the islands, as here on Mana. As sharply as they rise, they plunge down to a depth of some 90 metres (295 ft), creating dramatic drop-offs honeycombed with caves.

At the southern end of the chain, 89 islands have now been set aside as the Kornati National Park, to preserve their unspoilt nature, and their delicate, centuries-old balance between landscape and human activity. Visitors come on guided tours in boats, but this is also a treasured stopping point for yachts travelling along the Dalmatian coast.

The sea surrounding the park, protected since 1980 by restricted fishing, is rich is marine life of all kinds – fish, starfish, crustaceans, even bottlenose dolphins – which feed around swaying red and green algae and undulating fields of seagrass. The eroded holes and caves of limestone provide an inexhaustible playground for snorkellers and scuba-divers. They may also come upon the sunken quays built by the Romans.

The islands have been much more heavily populated in the past. Their history stretches back to Neolithic times, and they were occupied and ruled by the successive empires of the region: Illyrians, Greeks, Romans, Byzantines, Venetians. The islands offered a modest living through fishing, harvesting salt pans, raising livestock, beekeeping and growing olives.

They reached their greatest prosperity during the Venetian era, after 1524, when night-fishing using lights was introduced to catch sardines. Small villages were built to house the fishermen during the summer season, awaiting the precious moonless nights. The Venetians taxed this fishing, and built forts on the coast to regulate it, but the trade collapsed along with the Republic of Venice at the end of the 18th century.

The Kornati Islands have always served as a refuge from troubles on the mainland, but were also exposed to marauding pirates and the successive tides of empire. Now half of them are protected from the depredations of the modern world by the regulations of the National Park. Within this, four areas are completely out of bounds. Marked off as 'Strictly Protected Zones', they can be visited only by authorized scientists who monitor them to see what happens when nature is allowed to run its course in this harsh but fragile environment.

Most of the islands have low and smoothly rounded profiles, often hard to distinguish from distant clouds. Some robust varieties of trees can win a foothold in this landscape of arid limestone karst, but they rarely grow high enough to alter the skyline.

ILE DE RÉ

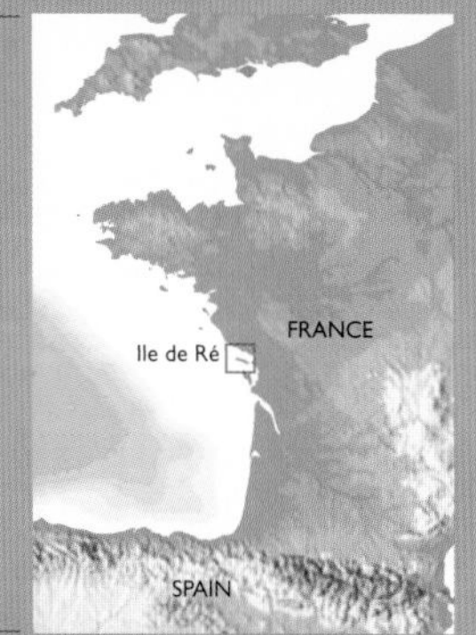

Latitude 46°12'N **Longitude** 1°26'W

Area 85 square kilometres (33 sq miles)

Status French (Department: Charente Maritime)

Population 17,000

Capital St-Martin-de-Ré

Official language French

Currency Euro

The first view of the Ile de Ré, from the elegantly curving road-bridge that now links it to the mainland at La Rochelle, is a flat plate of land heading out into the silvery Atlantic light. A single two-way road threads for 30 kilometres (19 miles) through this fluttering ribbon of an island, past vineyards, conifer woods and salt marshes, leading eventually to the Phare des Baleines, a lighthouse that looks down on the sandy crescent of the island's longest beach, the Plage de la Conche. Here, on a summer's day, the sea and sky turn porcelain blue. The thin line in the distance is the coast of the mainland, or

Le Continent as the islanders – the Rétais – call it, underlining their sense of disconnection from it.

The Ile de Ré is indeed a world apart, and it is considered by mainland France to be enviably chic. The permanent population of 17,000 clusters around the capital at St-Martin de Ré, and nine scattered villages. In and around them are thousands of holiday homes, with whitewashed walls – offset by hollyhocks, the island flower – and green shutters, beneath a roof of red ceramic tiles. Visitors come to the Ile de Ré all year, but in summer the holiday homes and campsites swell the population to some 200,000.

This is an island for family holidays. People come for the beaches and the watersports, and to mosey along the intricate network of cycle paths. At low tide, many head down to the shore with a bucket, net and winkling tools to collect shellfish and crabs – a traditional pursuit called *pêche à pied* (fishing on foot). The villages host high-quality markets, vibrant with colour, and packed with stalls selling gastronomic specialities.

The picturesque capital, St-Martin-de-Ré, is built around a harbour used by yachts and fishing boats. Restaurants lining the promenade cater for a busy, year-round flow of visitors.

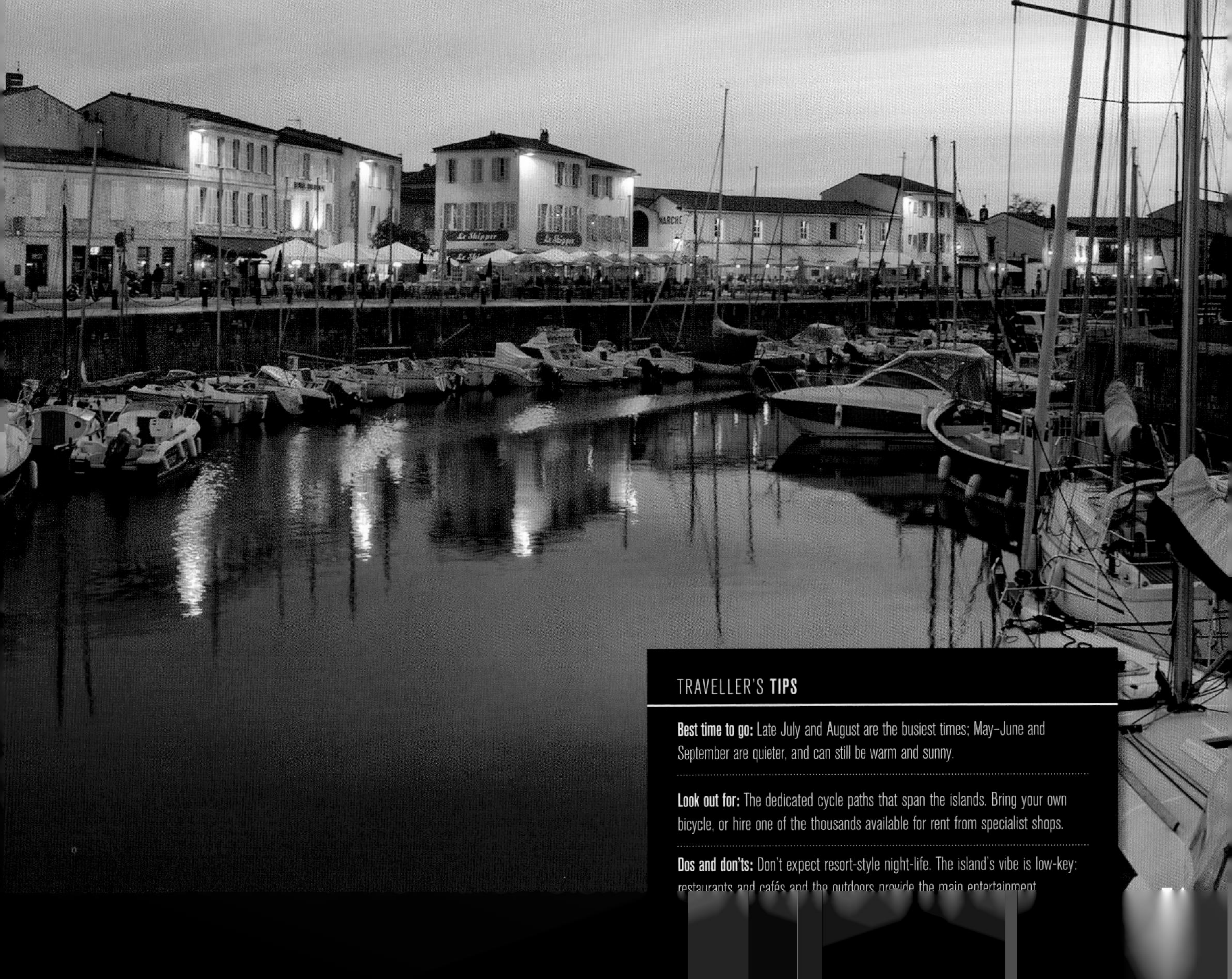

TRAVELLER'S **TIPS**

Best time to go: Late July and August are the busiest times; May–June and September are quieter, and can still be warm and sunny.

Look out for: The dedicated cycle paths that span the islands. Bring your own bicycle, or hire one of the thousands available for rent from specialist shops.

Dos and don'ts: Don't expect resort-style night-life. The island's vibe is low-key; restaurants and cafés and the outdoors provide the main entertainment.

Sea salt is harvested using centuries-old methods in the flat salt-marshes (*marais salants*) at the western end of the island. Sea water passes through sluice gates into ponds, where the water evaporates in the summer sun, becoming increasingly salty. In the final, shallow saltpans, it crystallizes into salt, which is raked into heaps to dry.

The Rétais thrive on the income that the visitors bring with them, but they also remain grounded in their traditional trades. They grow the grapes that make the island wine, and the fortified wine apéritif called Pineau des Charentes. The *ostréiculteurs* make a living by raising oysters. Others work the chequerboard salt pans in the marshlands that encircle the huge inlet called the Fier d'Ars. This salt used to be a mainstay of the island economy, transported by *ânes en culotte* – donkeys wearing leggings of chequered cloth to protect them from flies.

The low, flat terrain is explained by the island's origin as sandbanks: four islands now knitted by silt into one. Its

Straw bales, beside the ruins of the 12th-century Abbaye des Châteliers, are a reminder that the Ile de Ré has a strong agricultural base. Monks lived at the abbey until it fell victim to the Wars of Religion in 1623.

location on the shipping lanes into La Rochelle gave it an important strategic role, as witnessed by the splendid star-shaped fortifications – designed by Vauban in the 1680s – that surround the capital. Within their wide embrace to this day is one of France's top-security prisons: until 1938, this was the collection point for the *bagnards* – prisoners despatched on a one-way trip to the dreaded penal colonies of French Guiana. But St-Martin wears this history lightly. With pastel-shaded buildings surrounding its little harbour, and sloping streets lined with chic boutiques, St-Martin is outstandingly pretty, and a worthy capital to an island whose charms – even after a century of tourism – seldom fail to beguile.

CORSICA

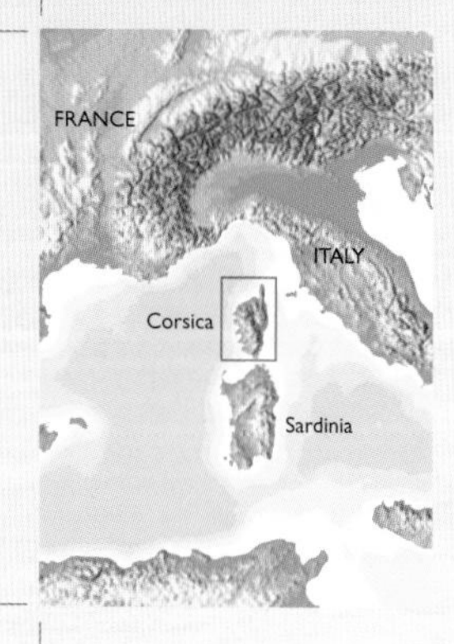

Latitude 42°9'N **Longitude** 9°5'E

Area 8,680 square kilometres (3,351 sq miles)

Status Region of France

Population 302,000

Capital Ajaccio

Official language French

Currency Euro

'Ile de Beauté' is the official nickname for Corsica. Island of Beauty is a hard label to live up to – but Corsica has little trouble in matching that promise. This is a spectacularly beautiful island.

Corsica is said to have the most varied landscape of all the Mediterranean islands. Most of the interior consists of high mountains covered with forests of pines, chestnuts and oaks, or by dry maquis scrubland populated by holm oaks, cork oaks and aromatic shrubs and plants such as laurel, thyme and wild mint. The Balagne region of the northwest is the 'Garden of Corsica', producing oranges and lemons, olives and olive oil, grapes and wine, and prickly pear fruits. To the south of the Golfe de Porto on the west coast there is an unworldly orange-and-pink landscape of weathered granite cliffs, pillars and spires, called Les Calanches (or Les Calanques in French), some soaring 300 metres (1,000 ft) above the sea – cherished by walkers as well as by scuba-divers.

The Désert des Agriates on the north coast, by contrast, is an uninhabited wasteland. A mountainous area in the northeast is famous for its chestnut forests, and is called La Castagniccia, from the Latin *castanea* (chestnut). When the Genoese rulers of Corsica purloined all the island's wheat in the 15th century,

TRAVELLER'S **TIPS**

Best time to go: May–June, for wild flowers and enough warmth for the beach. The island becomes very busy from mid-July to mid-August.

Look out for: There are many marked hiking trails all over the island, including the challenging GR20. Ask for details at the tourist offices.

Dos and don'ts: Do hire a car if you plan to do any touring. The local public transport, provided mainly by buses, is slow and infrequent.

Crowned by a 17th-century Genoese watchtower and lighthouse, the four islands of the Sanguinaires Archipelago point westwards from the Gulf of Ajaccio. Their name is said to be derived from the blood-red colours of the sunsets.

the Corsicans were forced to turn to flour made from chestnuts – still used in distinctive bread and pastry, and to feed pigs to produce cherished forms of pork and ham.

The Genoese, who ruled Corsica from 1347 to 1729, were just one of the many sets of external rulers that have laid claim to the island: Etruscans, Romans, Vandals and Visigoths, Franks, Saracens, Lombards, French – even the British, briefly, during the 1790s. In part this is because Corsica has always been curiously difficult to box neatly. It has been French since 1770, but it lies closer to Italy – only 11 kilometres (7 miles) from Sardinia, its much larger Italian twin to the south; the Côte d'Azur in France, on the other hand, is 170 kilometres (106 miles) to the north. The Corsican language is closer to Italian than to French: it is spoken by 65 per cent of the population, and signposts are written in both Corsican and French. A hero of Corsica is Pasquale Paoli, who led an independent Corsica from 1755 to 1769, with an assembly established in the inland fortress city of Corte. On the other hand, another famous Corsican, Napoleon Bonaparte, who was born in Ajaccio, was instrumental in nailing Corsica's colours to the French mast.

For the past 30 years or so, the French government has had to contend with the violent campaign for independence of the Corsican National Liberation Movement (FLNC). But even if they might sympathize with its sentiments regarding national identity, most Corsicans have shown their distaste for its extremism. For one thing, it harms Corsica's international reputation, and tourism is now a key part of the economy.

It is, above all, natural beauty that draws visitors to Corsica. One-third of the island is set aside as the Parc Naturel Régional de Corse, which covers the main central mountain ridge before plunging into a marine reserve at the rocky Scandola Peninsula in the northwest. And straddling the island is the famous GR20, the most challenging of France's Grande Randonnée long-distance walking paths, and one of its most spectacular. Called Frà Li Monti ('Between the Mountains'), it runs for 180 kilometres (110 miles) from a point close to Calvi in the northwest to the east coast just north of Porto Vecchio. Because of its high elevation, much of it more than 2,000 metres (6,500 ft) above sea level, the GR20 is passable only between the months of June and October, and it takes at least two weeks to complete. The Ile de Beauté challenges as much as she attracts.

Dawn casts its soft, rose-coloured light on the medieval citadel of Bonifacio, set strategically on a limestone promontory 70 metres (230 ft) above the sea and beaches. This port town lies close to the southernmost tip of Corsica: Sardinia is within sight, just 12 kilometres (7 miles) across the Strait of Bonifacio.

MONT-ST-MICHEL

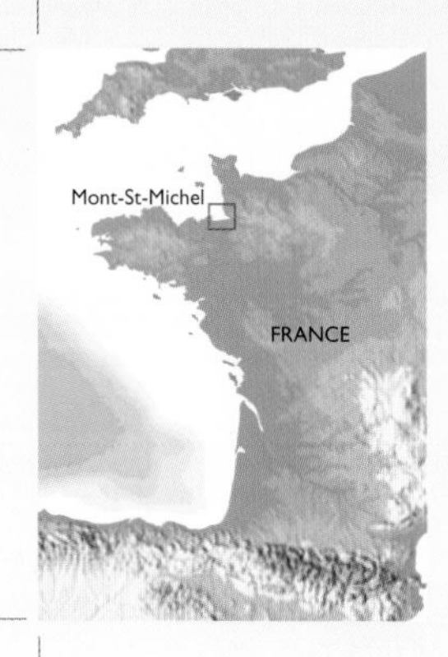

Latitude 63°18'N **Longitude** 20°36'W

Area 0.97 square kilometres (0.37 sq miles)

Status Island in Normandy, France

Population 40

Main settlement Mont-St-Michel

Official language French

Currency Euro

For miles around this crook in the coast of northwestern France, in the borderlands between Brittany and Normandy, tantalizing glimpses can be seen of Mont-St-Michel. It cuts an unmistakable silhouette, a fairytale citadel perched on a rough clump of rock, surrounded by sea, or, at low tide by undulating sheets of glistening mudflats.

This eye-catching profile drew thousands of pilgrims to their destination in medieval times, just as it draws some 3.5 million visitors every year now. The rock plinth had been sacred to the local Celts in pre-Roman times: this is where the souls of the dead resided, and it became known as Mount Tomb. But in the eighth century St Aubert, Bishop of Avranches, had a recurring dream that the archangel St Michael was calling him to build a shrine on the island, and in AD 708 St Aubert obliged.

The shrine developed into one of the most frequented pilgrimage sites of northern Europe. Benedictine monks settled here, and the Dukes of Normandy saw that the abbey church was developed in suitably grand Romanesque and then Gothic styles between the 11th and 13th centuries.

Meanwhile, by way of an earlier gift from Edward the Confessor, King of England, the Benedictines also owned St Michael's Mount in Cornwall, where a smaller version of the abbey was built.

Mont-St-Michel was heavily fortified, and was subject to occasional attacks. The English laid siege to it three times during the Hundred Years War (1357–1453). But, with the help of St Michael, it held out valiantly and became a symbol of French resistance. The 13th–15th century defensive walls – which remain remarkably complete – bear two cannon that are souvenirs of the English siege of 1434.

The abbey church rises to a pyramidal crescendo above the mud flats at low tide. At the base, the curtain of fortress walls provided a formidable defence against attack.

TRAVELLER'S **TIPS**

Best time to go: Any time of year. Bright weather lights up the architecture and the surrounding sea. The summer holiday period (July–August) can be uncomfortably busy.

Look out for: The gilded statue of St Michael slaying the dragon on the tip of the spire. Installed in 1897, it is by Emmanuel Frémiet.

Dos and don'ts: Don't walk out on the mudflats without a guide. The tide comes in very fast, traditionally said to be 'at the speed of a galloping horse'.

Restaurants, souvenir shops and hotels fill the steep Grande Rue, and occupy many of the 60 or so classified historic buildings. The crowds and commercialism may seem to conflict with the spirituality expected of such a sacred place, but similar scenes were no doubt played out by throngs of pilgrims in Mont-St-Michel's medieval heyday.

The fortunes of Mont-St-Michel declined during the Reformation, when pilgrimage went out of fashion, and by the 18th century it had been all but abandoned. The abbey was deconsecrated during the French Revolution and turned into a prison, until outrage – orchestrated by Victor Hugo, among others – over this tragic fate of a much-loved piece of French heritage shamed the government into action; the prison was closed and restorations began in 1863. Benedictine monks returned here in 1966, but they were replaced in 2001 by a new order of monks and nuns called the Monastic Fraternities of Jerusalem.

Pilgrims used to reach the island on a causeway that was exposed only at low tide. However, in 1879, a raised road was built above the reach of the high tide. This had a detrimental effect on the ecology of the bay, as silt from the inflowing River Couesnon could no longer be washed through and so clogged up the seabed. To rectify this, a project is underway involving the replacement of the causeway by a raised bridge for pedestrians, bicycles and shuttle buses, and the construction of a barrage that will help to ensure the flow of water around the island.

The refectory occupies one of the three floors of La Merveille ('The Marvel'), the Gothic wing on the north side of the abbey that provided the monks' living quarters and lodgings for pilgrims. Built in around 1212, it is one of the oldest parts of the monastery.

ELLESMERE ISLAND

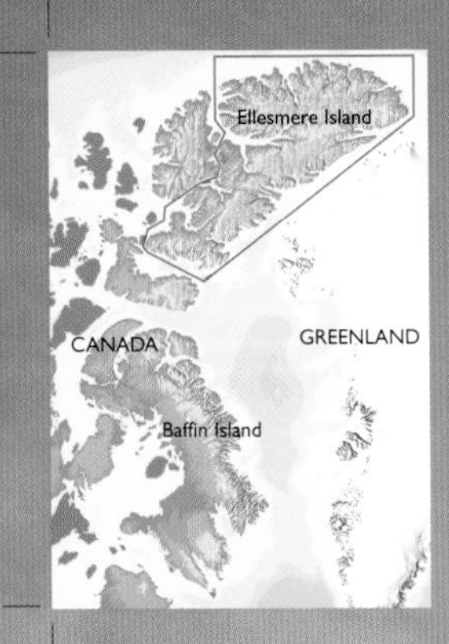

Latitude 79°50'N **Longitude** 78°0'W

Area 196,235 square kilometres (75,766 sq miles)

Status Island of the territory of Nunavut, Canada

Population 150

Main settlement Grise Fiord

Official languages Inuinnaqtun, English, French

Currency Canadian dollar

There is set of wooden huts at Fort Conger, built by the great American polar explorer Robert Peary. Returning here in 1899, after his first failed attempt to reach the North Pole in the winter darkness, with temperatures hitting -52°C (-62°F), he asked his loyal African-American companion Matthew Henson to help him off with his boots. Most of his toes, frostbitten, came off too. This impaired Peary's ability to walk, so when he and Henson made their fourth expedition in 1909, and claimed at last to be the first ever to see the North Pole, Peary was hauled most of the way on a dog sled.

Ellesmere Island is a place of extremes. It is the most northerly part of North America, and Fort Conger is close to the very top, Cape Columbia, which lies just 756 kilometres (470 miles) from the North Pole. Peary had to travel in winter, because that is when the island is bound to the North Pole by ice.

It is dark on Ellesmere Island from November to February; and it is light from May to August. The midnight sun is strangely disorienting, moving right around the horizon, casting shadows like a 24-hour sundial. The air is so pure and clear that distances are hard to judge: it seems as if you should be able to reach out and touch things that are in fact in the far distance. In summer the snow melts, exposing the land. Delicate wildflowers nod in the chill breezes: yellow arctic poppies, red moss campions and white mountain avens. In these cold temperatures there is little evaporation: but there is little precipitation either – just 60 millimetres (2½ in) a year, be it snow, rain or condensation. Ellesmere Island is technically a desert. The nearest trees are

Sculptural peaks surround Tanqueray Fiord in Ellesmere Island National Park. The island's many inlets and fjords are never entirely free of ice, even in the summer.

TRAVELLER'S **TIPS**

Best time to go: Late May to late August, when the average temperature is 7°C (45°F). Few visitors come in the winter.

Look out for: In the National Park and around Lake Hazen there are archaeological remains of campsites used by the first settlers 2,000–4,000 years ago.

Dos and don'ts: Unless you have survival training for wilderness travel, or experience of polar bear country, do hire the services of a guide.

Arctic hares are perfectly camouflaged for a snowy landscape populated by hungry hunters, such as Arctic wolves and Arctic foxes. They hibernate in the depth of winter by burrowing into the snow. In summer, however, their coats change to grey-brown to match the tundra. They feed on woody shrubs as well as berries and purple saxifrage.

Musk oxen are intensely social animals, living in groups of 12–60. Protected by their horns and bony foreheads, which they can use as vicious weapons of attack, they stand in protective rings around their calves, as a defence against predators, such as Arctic wolves. Musk oxen overwinter on Ellesmere Island, feeding on lichens scratched out of the snow, and insulated by their thick woolly coats.

2,000 kilometres (1,250 miles) to the south. In 1988 the pristine beauty of this environment was enshrined in the Ellesmere Island National Park, covering much of the far northeast of the island. It includes the largest lake in the Arctic Circle in North America, Lake Hazen, which in summer attracts nesting migratory birds such as the gyrfalcon, geese, plovers and mallards. Musk oxen and Arctic hares come to graze here, as well as Peary's caribou (reindeer) – smaller, more delicate and whiter than their relatives elsewhere in Canada.

This is an exclusive destination, reserved for the determined, with a considerable budget, who arrive here by small aircraft that land at the Tanquary Fiord Airport, or on an ice-breaking cruise ship. Although Ellesmere Island is twice the size of Iceland, it has a population of only 150, most of whom live in Grise Fiord, in the south of the island. This was founded by the Canadian government in 1953: prior to this no one had been living on the island for 200 years. Yet humans came here as long as 4,000 years ago, and the Thule people – ancestors of the modern Inuit – arrived 1,000 years ago.

Ellesmere Island was named by a British expedition in 1852, after the Earl of Ellesmere; today it is part of the Territory of northeastern Canada – established in 1999 – called Nunavut, meaning 'our land' in the language of the Inuit. And fittingly their name for Ellesmere Island National Park is Quttinirpaaq, meaning 'Top of the World'.

CUBA

Latitude 23°8'N **Longitude** 82°23'W

Area 109,886 square kilometres (42,427 sq miles)

Status Independent island republic

Population 11,239,400

Capital Havana

Official language Spanish

Currency Cuban peso

Walking through the streets of Havana, it is not hard to see what a grand city this used to be. Avenues of once-elegant villas and mansions converge on the centre where the former seat of the Cuban government, El Capitolio, replicates Washington's Capitol. Elaborate baroque churches recall a history going back to the Spanish conquistadores. The National Museum of Fine Arts of Havana has an astonishing collection of European masters: Brueghel, Rubens, Velázquez, Canaletto, Reynolds, Gainsborough, Goya, Turner, Constable.

Yet life has been put on hold here for generations. Murals, posters and souvenirs showing Fidel Castro and Che Guevara tell the story: since the Communist revolution that ousted the corrupt regime of Fulgencio Batista in 1959, Cuba has defied the sanctions of the USA, 145 kilometres (90 miles) to the north. For three decades Cuba was supported by the Soviet Union. When the USSR collapsed in 1991, Cuba's Communist government had to find a way to survive. The answer lay in tourism: not from the USA, which prohibited its citizens from travelling to Cuba, but from everywhere else.

One of the thousands of carefully preserved vintage American cars heads over a crossing in central Havana. Import restrictions have turned Cuba into wonderland for car enthusiasts.

TRAVELLER'S **TIPS**

Best time to go: March–April and September are best. December–February can be grey. The rainy season, with high humidity, lasts from May to October.

Look out for: There are masses of souvenirs (T-shirts, key-rings, wall-hangings) featuring Fidel Castro and Che Guevara, now freely used as fashion motifs.

Dos and don'ts: It is unwise to talk too freely about political issues. This is still a highly controlled society, with government ears everywhere.

Icon of the Cuban Revolution of 1959, Ernesto 'Che' Guevara appears in murals, posters, T-shirts and souvenirs. Born in Argentina, he had been involved in revolutionary politics in South America before meeting Fidel Castro in Mexico. Although he died in Bolivia in 1967, his image is still a symbol of Communist idealism in Cuba.

The elaborate baroque Cathedral of San Cristóbal in Havana was designed by an Italian for the Jesuits and built 1748–87. The city had been founded in 1515, just 23 years after Columbus had first reached the 'New World'.

Music is a living tradition in Cuba, the accompaniment to every social occasion. The American musician Ry Cooder recognized the neglected genius of Cuba's musical heritage, and brought it to world attention in the 1990s with his recordings and film of veteran musicians gathered under the name of the Buena Vista Social Club.

And Cuba is a gift for tourists. It's the largest island of the Caribbean, with great regional variety between the deep red tobacco-growing landscape of Viñales and Pinar del Rio in the northwest to Santiago, Cuba's second city, in the far southeast. In between there are rainforests, mountains, national parks, historic monuments and vestiges of the colonial past, notably the city of Trinidad. And all around are sandy beaches and idyllic outlying cays (small islands) lapped by the Atlantic to the north and the Caribbean to the south.

Varadero, on the narrow Hicacos Peninsula on the north coast, is a special tourist enclave. In the 1930s, this was a favourite hideaway of Hollywood filmstars, industrial tycoons and the Chicago Mafia boss Al Capone. Now the white-sand beach, 20 kilometres (12 miles) long, is lined with modern, top-quality hotels. It is an idealized version of the Caribbean island with beach and blue seas, buffet dinners featuring fresh lobster and *la cocina criolla* (the Spanish-African blend of cooking), along with rum daiquiris and *mojitos* (flavoured with fresh mint) – plus the world's finest cigars. Visiting performers bring dazzling shows featuring traditional Cuban dances, such as the high-octane rumba, and the more stately *habanera* and *bolero*.

And there is music everywhere, as there is right across the island. Small four- or five-piece bands, with guitars, African percussion and the distinctive clarion call of Cuban trumpet, play traditional *son* – rhythmic, soulful songs about life and love. You can never be far from a rendition of *Guajira Guantanamera* ('The Girl from Guantanamo'), composed in 1928, and now virtually the national anthem.

ARUBA

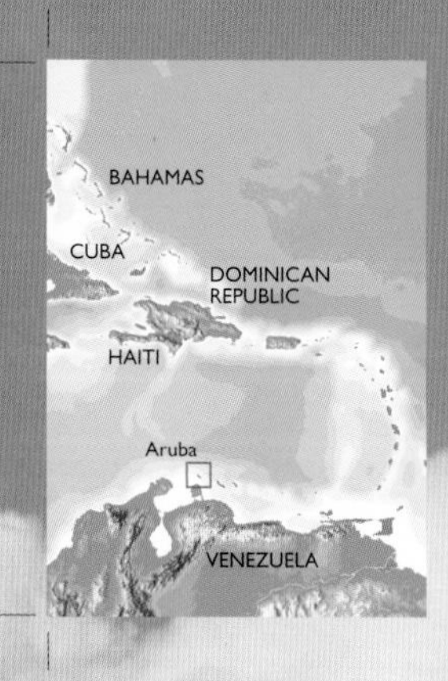

Latitude 12°31'N **Longitude** 70°1'W

Area 180 square kilometres (69 sq miles)

Status Part of the Kingdom of the Netherlands

Population 103,000

Capital Oranjestad

Official languages Dutch, Papiamento

Currency Aruban florin

Aruba is the kind of island that owes its modern prosperity to the beach-life dream. White-sand beaches line up along the south and west coasts, adjacent to comfortable hotels and luxury resorts. The sea is blue, the sand soft, and the sunshine – moderated by sea breezes – delivers a more-or-less constant temperature of around 28°C (82°F) throughout the year.

It has not always been so appreciated. When the Spanish conquistadores toured the region they labelled it an *isla inútil* (useless island) because it had no gold and little water.

A divi-divi tree strikes a balletic pose in its unlikely setting: growing on the beach. These trees, *Caesalpinia coriaria*, are found throughout the Caribbean, but they are especially treasured in Aruba for their beautifully gnarled, sculptural shapes.

TRAVELLER'S **TIPS**

Best time to go: Any time of the year is good. What little rainfall Aruba has comes mainly at night. The high season is January–March.

Look out for: The walk up Hooiberg (Haystack Mountain): its 562 steps reach a height of 168 metres (541 ft), offering panoramic views as far as Venezuela.

Dos and don'ts: Don't expect to see the much-photographed Natural Bridge on the north coast: it collapsed in 2005. The sea has carved out other smaller ones.

Amerindians, however, had made it their home for some 3,000 years: their cave paintings can be seen in Fontein Cave in the Arikok National Park. In around AD 1000 Caquetio Arawak Indians settled on the island, seeking refuge from marauding Caribs. In 1513 the Spaniards enslaved them and shipped them off to Hispaniola, but some returned and – proving a gift for handling horses and cattle – took up ranching for the Spanish and, after 1636, for the Dutch. The Dutch brought in Asian labour and some African slaves – creating a broad ethnic mix that characterizes the Aruban population today.

Aruba is one of the ABC islands – Aruba, Bonaire and Curaçao – that lie close to the Venezuelan coast and share a Dutch history. Aruba is actually a constituent part of the Kingdom of the Netherlands: its people are Dutch citizens and the head of State is the Dutch monarch. This makes it a dream Caribbean island for the Dutch in particular. They come here to find a mixture of the exotic and the familiar: the streets of Oranjestad, the capital, are lined with Dutch-style buildings, with step-gabled façades – but painted in brilliant Caribbean colours to square up to the tropical sun.

ST BARTHELEMY

Latitude 17°54'N **Longitude** 62°50'W

Area 21 square kilometres (8 sq miles)

Status Overseas territory of France

Population 8,800

Main settlement Gustavia

Official language French

Currency Euro

There is always a sense of privilege to fly into an island on a tiny turboprop aeroplane. The airport at St-Jean is not long enough to take larger planes, so this is how visitors arrive by air, mostly after a ten-minute hop from the international airport on neighbouring St-Maarten. Others arrive by private yacht, coming into the shelter of Gustavia's harbour, or anchoring out in the turquoise bays. The facilities are too small for cruise ships. And that is how St Barthélemy likes it: privileged and exclusive – for this is one of the world's most glamorous and fashionable holiday destinations. The international rich and famous come here, to stay in the choice boutique hotels or in the many luxury villas that dot the hills. The peak of the winter high season is New Year's Eve, when the harbour fills with massive private cruisers, and heads pop in the street to spot the stars of stage and screen.

When Christopher Columbus found the island in 1493, and named it after his brother Bartolomeo, it seemed unpromising: a rocky volcanic island encircled by shallow lagoons and white-sand beaches. With steep hills and no springs, it was not suitable for plantations. But the French laid claim to the island in 1648, and placed settlers from Brittany and Normandy on it to farm and fish. The result is a slice of the Côte d'Azur in the Caribbean, and in the off-season it reveals its more modest roots, as it was until the 1970s, before the glamour swept in.

Colourful chalets line the edge of Flamands Beach. All the buildings here and elsewhere on St Barts are low-rise: by law, they cannot be taller than a palm tree.

TRAVELLER'S **TIPS**

Best time to go: High season for the super-rich runs from 15 December to 15 January. For low-season prices, go in April–August.

Look out for: Celebrities. A number of filmstars and rockstars have villas on the islands, and others come to rent them.

Dos and don'ts: Do hire a car or jeep. It is a small island, but there is plenty to explore and discover: villages, beaches, panoramic views from the hills.

MARTINIQUE

Latitude 14°40'N **Longitude** 61°00'W

Area 1,128 square kilometres (435 sq miles)

Status Overseas department (and region) of France

Population 398,000

Capital Fort-de-France

Official language French

Currency Euro

The ragged northern hills of Martinique are carpeted with tropical forest – a deep forest of towering trees. Parrots call out from high in the canopy; the ground is swathed in dense undergrowth. Distant surf can be heard crashing against remote coves. At nightfall, tiny tree frogs chant and whistle. And when it rains – as it does frequently – drips from above drum on the broad leaves to create an all-enveloping hum. The humid, peaty scent of leaf compost rises from the ground. Every shade of green is here – the expression of exuberant fertility. This same impression is rolled out to almost every part of the island, even

TRAVELLER'S **TIPS**

Best time to go: The high season coincides with the 'dry' (or at least drier) season lasting from February to May. But the climate is generally agreeable.

Look out for: Martinique's rum, called 'rhum agricole', has a distinctive farmyard scent. The classic ti-punch is a mixture of white rum, cane syrup and lime juice.

Dos and don'ts: Do visit the Jardin de Balata, a private tropical garden with trees, palms, giant ferns, flowering plants and a suspended walkway over the canopy.

where the land has been tamed by agriculture – in the banana plantations, the fields of sugar cane, in the curving coconut palms that lean over the beaches. In the towns and villages, though, the greenery is punctuated with wild splashes of colour: flame trees, hibiscus, canna lilies, drifts of bougainvillea – all lit by a bright tropical sun.

At Le Carbet, Anse Noire and elsewhere, black-sand beaches – vividly contrasting the bright paintwork of fishing boats drawn up on them – are stark reminders of the root cause of all this fertility: the rich volcanic soil. Volcanic activity created the island's jagged contours of hills and bays, around which the roads follow their serpentine courses, revealing new landscapes at every turn. In the far northwest, the island rises to a crescendo at the volcanic peak of Mont Pelée. It is a name that still produces shudders in the collective memory of the Martiniquais. In 1902 Mont Pelée erupted, and coughed up an avalanche of burning, poisonous gas and debris that poured down the mountainside and extinguished the capital city of St-Pierre. At the time, St-Pierre was fêted as 'the Paris of the Caribbean'. No such epithet is used

Although its flanks have been softened by vegetation and it is more than a century since its last devastating eruption, Mont Pelée still has a raw and brooding presence.

Fishing remains a major pursuit, often conducted on a non-commercial basis, such as taking fishing boats out at night to bring back a catch for family and friends. Seine fishing is also practised, with teams of up to 20 people taking a hoop of net perhaps 600 metres (1,968 ft) long out to sea, then hauling the catch to the shore.

of the new capital, Fort-de-France, a busy port and administrative centre of haphazard composition. Parisian or not, it is the capital of an island that remains resolutely French. Indeed, Martinique is not just a French island, it is actually a *département* of France – as French as the Ardèche or Dordogne, and part of the European Union. It sends deputies and senators to the French parliament; many of its students go to university in France; its shops stock familiar French brands.

And yet Martinique will always have its own, distinctive slant. The local language is a creole, only loosely attached to standard French. The stately traditional dances, such as the quadrille, waltz and mazurka – performed by women in long dresses and knotted headdresses of madras cloth – may be European in

Brightly painted fishing boats rest on a black-sand beach at Grand'Rivière, on the northernmost tip of the island. The boats are personalized with stencil-painted names.

origin, but they have an African clip and rhythm. The churches are mainly Roman Catholic, but the exuberant Lenten carnival processions are unquestionably Caribbean. The cooking is influenced by French cuisine, placing it among the very best of the Caribbean – but is tied to local produce, such as langouste (spiny lobster), land-crab, chilli, coconut, pineapple and papaya.

This Frenchness sets Martinique apart from the majority of islands in the Caribbean, and the Martiniquais are fiercely proud of the unique combination of forces that have shaped them. But this of course also entails the darker story of slavery, which resonates like a deep bass-note in Martiniquais culture. The poetry and music of the island, even the architecture, bear witness to this past. It is a subject addressed full on by La Savane des Esclaves, at Trois-Ilets, a reconstructed slave village that sensitively demonstrates the lives and work of its habitants. There is more to Martinique than beaches and sea, suggests its publicity. Indeed there is.

GUADELOUPE

Latitude 16°15'N **Longitude** 61°35'W

Area 1,628 square kilometres (629 sq miles)

Status Overseas department (and region) of France

Population 406,000

Capital Basse-Terre

Official language French

Currency Euro

From heaven to hell: a tranquil beach on the north coast of Grande-Terre lies at the base of the lagoon formed behind a ferocious inlet from the Atlantic, flanked by high cliffs known as the Porte d'Enfer (Hell's Gate).

Like Martinique, the island of Guadeloupe is a part of France. It is a French *département*, and so a distant outpost of the European Union. Every French citizen of *la métropole* (mainland France) grows up with the feeling that – out there, in the warm waters of the Caribbean – lies a warm and welcoming fragment of their nation, a place that beckons. The existence of Guadeloupe feeds some deep longing in the French national psyche. And sure enough, many French people will, at some point in their lives, head off on a dream holiday in the French Lesser Antilles.

Guadeloupe fully acknowledges that it represents the European vision of a tropical paradise, and does everything both to promote it, and to fulfil it. This is a ravishingly beautiful island – everything that a Caribbean island should be. There are white-sand beaches and brightly painted beach

bars, shaded with coconut palms that rustle in the salty *alizés* – the steady trade winds that keep the ambient temperature on just the right side of hot. A wild volcanic landscape forms the dramatic backdrop to the shoreline. The soil of Guadeloupe is so fertile you can almost see plants grow. The landscape is thickly draped with huge and wild specimens of the kind of plants that are vaguely familiar to visitors because they have seen them in botanical-garden hothouses, or as carefully tended houseplants on their own windowsills back home.

Guadeloupe is proud of its history, and the blending – or *métissage* – of the many different races that have been brought together by it: French, African, Indian, Chinese. It is a culture, as they say, that reveals itself in the streets rather than in museums – in the faces of the people, in the music and the folk art, in the local creole language, and in the spicy flavours of the cooking. Nonetheless, there are always reminders of metropolitan France here: from the *baguettes* in the shopping basket to the uniform of the *gendarmes* and the *mairies* and *tricolore* flags in every town. The name, however, is Spanish in its origin: Christopher Columbus arrived here in 1493, and called the island Santa María de Guadalupe, after a miraculous image venerated in a monastery in Guadalupe,

TRAVELLER'S **TIPS**

Best time to go: November–May is the high season, which coincides with the 'dry' (or at least drier) season. The wetter (also hurricane) season is June–November.

Look out for: Beauport, Le Pays de la Canne (Land of Sugar Cane), Grande-Terre: this old sugar refinery is a museum of the sugar industry and the slavery that sustained it.

Dos and don'ts: Be sure to walk in the National Park on Basse-Terre. It is a rainforest experience that you will never forget.

Extremadura, Spain. The French settled the island in 1635, and annexed it in 1674.

Guadeloupe and Martinique are often spoken about in the same sentence, but each has its special characteristics, just as the Dordogne differs from neighbouring Lot-et-Garonne back in *la métropole*. The two islands are in fact separated by some 130 kilometres (80 miles) of sea, and the English-speaking island of Dominica lies between them. They are two very distinct locations.

And Guadeloupe itself is not one island but several. There are two principal islands, called Grande-Terre and Basse-Terre, divided by a narrow channel. Confusingly Basse-Terre is actually bigger than Grande-Terre; together they form a butterfly shape, giving rise to the name Ile Papillon, but Grande-Terre also looks remarkably like a hummingbird, one of the most endearing of the islands' many bird species.

Basse-Terre is dominated by its massive volcano, called La Grande Soufrière, rising to 1,467 metres (4,813 ft). It is still active – the name refers to its sulphurous emissions. It is the centrepiece of a magnificent Parc National de la Guadeloupe, famed for its rainforest walks and waterfalls, and home to the rare and endangered lesser Antillean iguana and Guadeloupe racoon. The park also embraces the Réserve Naturelle du Grand Cul-de-Sac Marin, a bay formed by the northern crook between Grande-Terre and Basse-Terre, and an irresistible draw for scuba-divers.

Administratively, the name Guadeloupe also covers the islands of La Désirade and Marie-Galante, and the collection of eight little islands called the Iles des Saintes. Each has its own character, and level of tranquillity enforced by the consecutive stages needed to reach them. And each one delivers its own kind of magic.

The rocky cliffs of the Pointe des Châteaux on Grande-Terre, one of the main islands of Guadeloupe, jut out into the sea. On a clear day, it is possible to spot Antigua (43 miles away) and the Montserrat Islands (50 miles away) from some of the cliffs.

The outstanding Plage de la Ferrière is on the satellite island of Marie-Galante. Named by Christopher Columbus after his flagship on his second voyage, Marie-Galante was dubbed 'Island of 100 Mills' because of its many windmills, and ox- and steam-powered mills built in the 18th–19th centuries to process sugar cane in rum distilleries.

ST LUCIA

Latitude 14°1'N **Longitude** 60°59'W

Area 616 square kilometres (238 sq miles)

Status Independent island state

Population 174,000

Main settlement Castries

Official language English

Currency East Caribbean dollar

There is a strange and unworldly scene in the hills above the west coast of St Lucia: a tortured landscape of livid pink and grey, with fudge-like encrustations around steaming pits of bubbling mud. For miles around the air smells intriguingly of rotten eggs. Below, where the outflow has cooled, visitors slap mud onto their bodies, then wash it off in the volcanic water – a curiously elating experience.

The Sulphur Springs is just one of the many attractions of this Caribbean island. There are superb beaches to laze on, lapped

TRAVELLER'S **TIPS**

Best time to go: The high season is from January to April. May–November is the rainy season, but that usually means brief showers (and possibly hurricanes).

Look out for: The St Lucia Jazz Festival in early May, which attracts big international names. Events take place at venues all over the island.

Dos and don'ts: Do go to one of the public parties held on Fridays and Saturdays, such as the 'Jump-Up' at Gros Islet, for music, dance, local food and rum

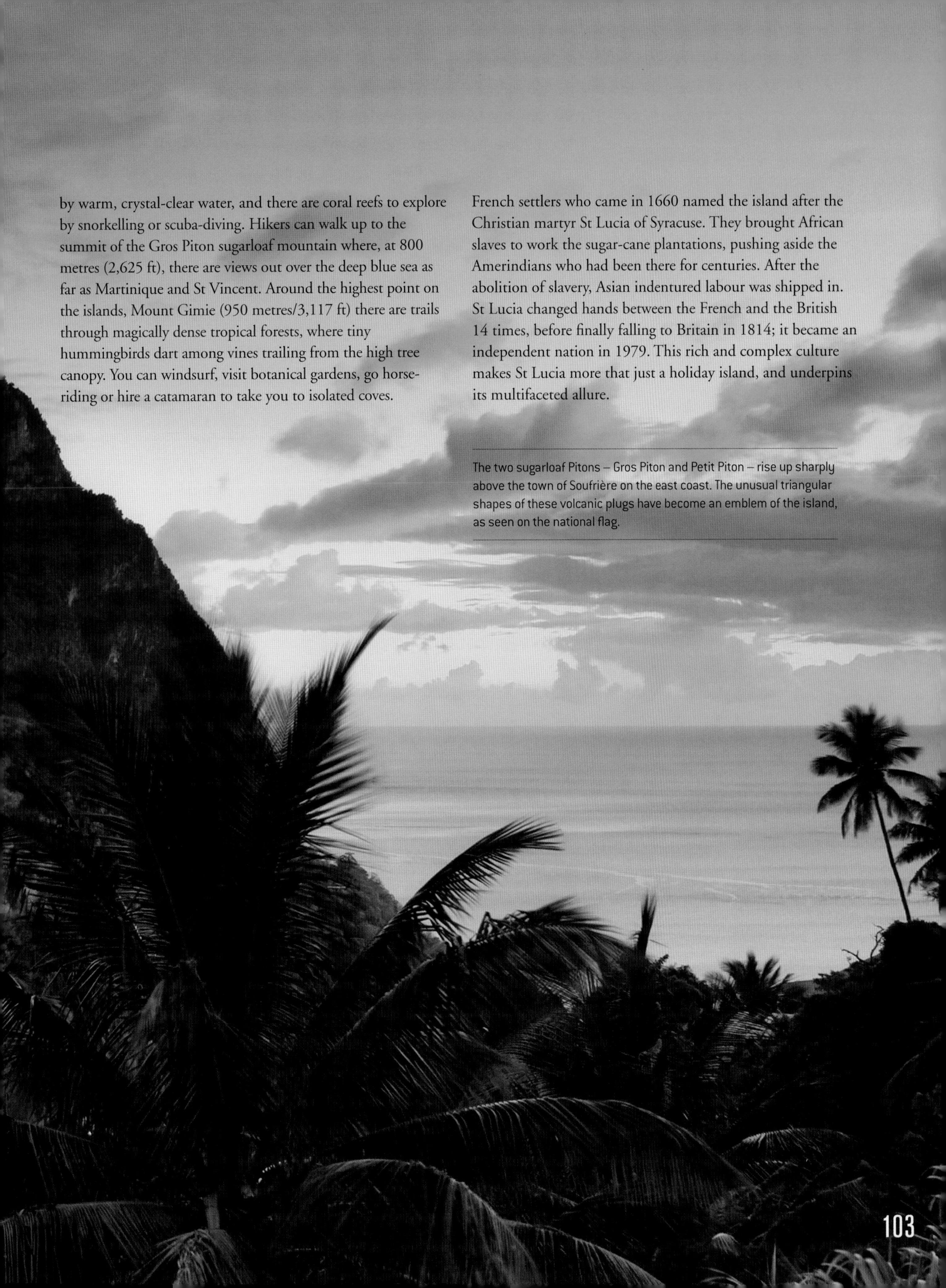

by warm, crystal-clear water, and there are coral reefs to explore by snorkelling or scuba-diving. Hikers can walk up to the summit of the Gros Piton sugarloaf mountain where, at 800 metres (2,625 ft), there are views out over the deep blue sea as far as Martinique and St Vincent. Around the highest point on the islands, Mount Gimie (950 metres/3,117 ft) there are trails through magically dense tropical forests, where tiny hummingbirds dart among vines trailing from the high tree canopy. You can windsurf, visit botanical gardens, go horse-riding or hire a catamaran to take you to isolated coves.

French settlers who came in 1660 named the island after the Christian martyr St Lucia of Syracuse. They brought African slaves to work the sugar-cane plantations, pushing aside the Amerindians who had been there for centuries. After the abolition of slavery, Asian indentured labour was shipped in. St Lucia changed hands between the French and the British 14 times, before finally falling to Britain in 1814; it became an independent nation in 1979. This rich and complex culture makes St Lucia more that just a holiday island, and underpins its multifaceted allure.

The two sugarloaf Pitons – Gros Piton and Petit Piton – rise up sharply above the town of Soufrière on the east coast. The unusual triangular shapes of these volcanic plugs have become an emblem of the island, as seen on the national flag.

TIERRA DEL FUEGO

Latitude 54°47'S **Longitude** 68°20'W

Area 73,753 square kilometres (28,476 sq miles)

Status Shared between Argentina and Chile

Population 126,000

Main settlement Ushuaia (Argentina), Porvenir (Chile)

Official language Spanish

Currency Argentinian peso/Chilean peso

A land of savage, pristine beauty, Tierra del Fuego sits like the flick of a wrist at the very base of South America. Here the Andes finally peter out. Having run down the entire course of the continent, they splinter into hundreds of tiny islands and one large one: Isla Grande de Tierra del Fuego.

It is a kind of world's end, and on cold days of glowering clouds and bitter winds, it can seem forsaken. Next stop Antarctica, and looking out towards it is Cape Horn, a name that brings shivers to any mariner. Clippers bringing tea from

China in the 19th century would have to run the gauntlet of its mountainous seas, with sailors chipping ice off the masts to adjust the sails.

Cape Horn sounds like a description of the shape of the land, the pointed tip of the continent. In fact it is named after Dutch town Hoorn, birthplace of Captain Willem Schouten who passed this way in 1616. Nothing is quite what it seems here. The name Tierra del Fuego, 'Land of Fire', suggests that the towering mountain peaks are seething volcanoes; they are

TRAVELLER'S **TIPS**

Best time to go: The warmest months are November to mid-March, but the marine climate keeps temperatures manageable in winter.

Look out for: The End of the World Train is the world's southernmost railway: a historic steam train travels from a point near Ushuaia into the National Park.

Dos and don'ts: Don't forget to take foul-weather clothing, whenever you go. Do remember that in winter the wind-chill factor can increase the effect of the cold.

Garibaldi Glacier marches slowly towards its terminus in the Garibaldi Fjord at the foot of Mount Darwin, on the Chilean side of Isla Grande.

Tierra del Fuego is one of the few places in the world where King Penguins live and breed. They do not build nests: instead, after the female has laid her single egg, both male and the female incubate it – in shifts lasting between two and three weeks each, for a total of nearly two months – by holding it on top of their feet.

not. The region was named by the Portuguese navigator Ferdinand Magellan as he led the first-ever European expedition through the islands in 1520 to reach the Pacific (and eventually to circumnavigate the globe, without him). As he passed through the strait that now bears his name he saw scores of fires in the landscape, lit by the inhabitants.

Those inhabitants astounded all early European visitors – including the great British naturalist Charles Darwin, on his journey to the Galápagos Islands in 1832. The Yaghan people lived by hunting and fishing, and went about completely naked, warmed only by animal grease and their fires. It now seems that their ancestors had been living in this region for 10,000 years: in other words, in the great early migrations from Asia through the Bering Strait in the far north, which brought the native peoples of the Americas, they were the ones that came to the farthest point south. But pioneering European settlers in the 19th century, taking the land for sheep and cattle farming, panning for gold and logging, brought disease and conflict to the native inhabitants. There were once perhaps 6,000 Yaghan and Selk'nam, who lived in the north of the Isla Grande; perhaps 1,600 Yaghan survive today, and no Selk'nam: they were actively wiped out by ranchers.

Conflict also explains the bizarre map of Tierra del Fuego. It is divided between Chile and Argentina, with an L-shaped border between them. From the north, this cuts across the eastern end of the Strait of Magellan, slices down the middle of the Isla Grande, then takes a sharp turn to the east along the Beagle Channel (named after ship that Darwin sailed on). All to the west and south of this line is Chile, including Isla Navarino to the south of Isla Grande, and Cape Horn.

But the land itself does not respect such borders. A majestic emptiness spreads right across the region. To the north is a softer landscape of pampas and peatbogs; further south, forests of deciduous beech trees rise past glacier-fed lakes to the belt of mountains that reach their apogee at Mount Darwin (2,438 metres, 7,999 ft) on the Chilean side. This is a landscape that humans can barely dent. Instead they must share it on even terms with a set of animals that have honed themselves to the conditions: condors, eagles, austral parakeets, guanacos (wild relatives of llamas), otters, seals, sea lions, penguins, albatrosses. Both countries recognize the preciousness of these islands and have established national parks: the Tierra del Fuego National Park in Argentina, and the Alberto de Agostini National Park in Chile. This is also the appeal of Tierra de Fuego to increasing numbers of visitors who come to fish and trek, sail and go horseriding in summer – when days have 17 hours of light – and ski and snowboard and go dog-sledding in winter. And, above all, to enjoy the rare sense of untrammelled space.

This lighthouse stands on the most northerly of five islets called Les Eclaireurs, in the Beagle Channel. In Argentina it is known as 'The Lighthouse at the End of the World'.

NANTUCKET, MARTHA'S VINEYARD

Latitude 41°20'N **Longitude** 70°18'W

Area Nantucket: 272 square kilometres (105 sq miles); Martha's Vineyard: 226 square kilometres (87 sq miles)

Status Islands of Massachusetts, USA

Population Nantucket: 10,000; Martha's Vineyard: 15,000

Main settlement Nantucket; Vineyard Haven (Martha's Vineyard)

The Cape Cod Peninsula in Massachusetts – with its clapboard houses, white picket fences, silvery Atlantic air and russet autumn colours – already sets hearts racing. Take one short step over the ocean to Martha's Vineyard, and another to Nantucket, and you are among dream islands.

Two aspects inspire these dreams. The first is that these islands are redolent with American history, and a particular aspect of it that speaks of independence, intrepid seamanship and the heroic fortune-making endeavour that accompanied the great

age of whaling in the 18th and 19th centuries. Both these islands, and especially Nantucket, were at the heart of the whaling industry – the world portrayed by Herman Melville in his *Moby Dick*, published in 1851. Whale oil, and spermaceti wax used to make candles, were valuable resources in the days before kerosene and petroleum, and created fortunes for the whaling captains, whose mansions can still be seen on both islands. The Whaling Museum in Nantucket vividly recounts this extraordinary story, which took sailors from these islands to the far-flung shores of the South Pacific and Antarctica.

TRAVELLER'S **TIPS**

Best time to go: July–August is the high season, and very busy. June and September can be more agreeable. But all seasons have their merits.

Look out for: Local seafood: tuna, swordfish, oysters, blue-shelled crab, Nantucket Bay scallops and clams (and particularly clam chowder).

Dos and don'ts: Do hire a bicycle and tour the islands on cycle paths. Nantucket is flatter than the Vineyard, but you may have to contend with stiff sea breezes.

In Nantucket Harbor, traditionally styled houses are built over or overlooking the water. Many are rented as holiday accommodation.

The second source of dream inspiration is the great charm and beauty of these islands, which make them perfect holiday destinations. Both the islands are sprinkled with stardust too: filmstars, famous writers and musicians, and US presidents come here to enjoy the tranquillity of the landscape and villages, the Old World courtesy and grace of the residents, and the sense of unflashy relaxation that comes with holiday islands at ease with themselves. Martha's Vineyard in particular attracts celebrities, as visitors and as second-home owners; Nantucket – one step more remote and quieter, and despite the regular arrival of the fog that has leant the island its nickname 'The Grey Lady' – has a roll call of mega-rich financiers and captains of industry. Real estate begins at millions of dollars.

But this does not make these islands exclusive. There are simple homes too, owned by islanders who have been here for generations, and there is a modesty at the heart of their culture, engendered by the religious roots of the founding fathers. Of the six towns of Martha's Vineyard, only two are permitted to serve alcohol. Nantucket – founded by Quakers – is proud to have no fast-food restaurants, no neon signs and no traffic lights. Every summer holiday season the population of the Vineyard (as it is known) multiplies ten times, and Nantucket's five times. Many visitors arrive by ferry from Cape Cod, a journey of about 11 kilometres (7 miles) for the Vineyard, 48 kilometres (30 miles) for Nantucket. They head for the beaches, they kayak and kitesurf, go shore-fishing and sailing, cycling and horseriding, tour the historical sites and

museums, eat excellent local seafood and hang out in the friendly cafés, bars and clubs. For an excursion, some cross on the ferry from Edgartown to the islet of Chappaquiddick, to immerse themselves in the tranquillity of the Cape Poge Wildlife Refuge and the My Toi Japanese Garden.

And who was Martha exactly? The origin of the name is obscure: it seems that in 1602 an Englishman called Bartholomew Gosnold saw some wild vines on another island and attached them to the name of his daughter Martha, and the name was later ascribed to this larger island. As for Nantucket, this was the name given to it by the Wampanoag Indians. Their affection for these islands dates back many thousands of years.

There is an old-world charm to Nantucket, as witnessed by its Main Street, with brick pavements and painted shop signboards. Nantucket was first settled by the English in 1659, but most of the historical buildings date from the 19th century – many after the Great Fire of 1846, which destroyed the wharves and much of the business district.

Great Point Lighthouse guards Nantucket's northernmost tip. This is a replica: the original stone tower, built in 1817, was destroyed in a storm in 1984 – a testament to the extreme weather that occasionally visits.

FLORIDA KEYS

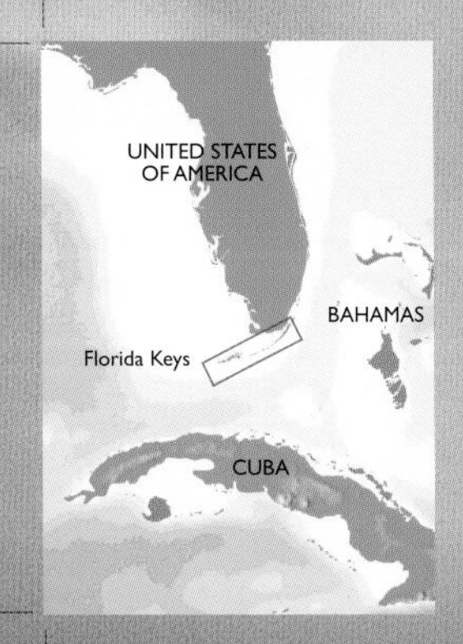

Latitude 24°40'N **Longitude** 81°32'W

Area 356 square kilometres (137 sq miles)

Status Islands of Florida, USA

Population 80,000

Main city Key West

Official language English

Currency US dollar

Florida Keys – a dream vacation: that is how the brochures put it, and that is how most Americans respond when they hear that name.

The Florida Keys form an extraordinary arc of about 1,000 coral and limestone islands, sweeping all the way from Elliott Key in the Biscayne National Park, 40 kilometres (25 miles) south of Miami, around the southern tip of Florida to the Dry Tortugas, some 300 kilometres (186 miles) away. The word Key comes from the Spanish *cayo*, small island – the Spanish

Many of the outlying islands, off the Overseas Highway, are uninhabited. Groups of these are now protected as nature reserves.

explorer Ponce de León first laid claim to these islands in 1513, and they remained in Spanish hands until 1819.

From the Florida mainland and Key Largo all the way to the main city, Key West, the islands are connected by the Overseas Highway, 204 kilometres (127 miles) long – the last and most spectacular leg of US Highway 1 on its journey down the East Coast from Maine. The drive along the Overseas Highway – with its 43 bridges, including the much-photographed Seven Mile Bridge – is reason enough to be here. Along the way is a chain of resorts, hotels, campsites, coffee shops and beaches – and, both to the left and the right, vast views of blue seas under white and mountainous cloudscapes.

The Florida Keys are a place of pilgrimage for divers and fishermen. Key Largo, at the northern end of the chain, calls itself 'the diving capital of the world', with reefs and wrecks, and ships deliberately sunk. The John Pennekamp Coral Reef State Park covers 240 square kilometres (93 sq miles), and contains a wonderland of marine life that can be enjoyed not

TRAVELLER'S **TIPS**

Best time to go: The high season lasts from mid-December to just after Easter. Note that hurricanes may occur any time between June and November.

Look out for: Key deer, a species of miniature deer found only here; they can be seen at the National Key Deer Refuge on Big Pine Key and No Name Key.

Dos and don'ts: Do book your accommodation well in advance during the high season, particularly for Fridays and Saturdays, and especially for Key West.

only by scuba-divers and snorkellers, but also by non-swimmers in glass-bottom boats.

The set of islands grouped together as Islamorada is famous as a centre for sport fishing. Boats go out in search of blue marlin, sailfish, shark, tuna, mahi mahi and tarpon. Anglers fish from the shore, others in the mangrove swamps, others still from the piers and bridges. The end of the road is Key West, the largest city by far in the chain, with a population of 25,000. It's a breezy fun-loving place, with the raffish air of a frontier outpost. The Old Town has wooden clapboard houses and elaborate Victorian gingerbread mansions, redolent of the decades in the 19th century when this was the largest city in Florida.

Among the city's other claims to fame are the fact that Ernest Hemingway made his home here for a decade. He wrote some of his most famous works – notably *A Farewell to Arms* – in the comfortable house on Simonton Street that is now his museum. When he wasn't writing, he spent a good deal of his time in Sloppy Joe's bar, which is still the focal point of Key West's main drag, Duval Street. He was sitting here one evening, after a happy day's fishing, when Martha Gellhorn walked in off the street. She became one of the great loves of his life, as well as his lifelong rival as a writer and war reporter.

Key West developed as a salvage centre, dismantling ships that came to grief in the treacherous surrounding waters. It was also a trading port, with links between New Orleans and Cuba, which lies just 145 kilometres (90 miles) south, and to the Bahamas, 220 kilometres (135 miles) to the east. In 1912, through an extraordinary feat of engineering, a rail connection was forged all the way along the archipelago to Key West; but the line was wrecked in the Labor Day Hurricane of 1935, one of the worst ever to hit the USA. The State of Florida subsequently bought all the surviving structure and used it to create the Overseas Highway, but the remnants of the railway are still to be seen along the way, and they have all the redundant grandeur of a Greek temple or Roman aqueduct.

One more extraordinary feat of endeavour lies beyond Key West. At the very end of the archipelago are the uninhabited Dry Tortugas. This is the site of the largest brick structure ever built in the Americas: the vast hexagonal Fort Jefferson. Begun in 1846, it was occupied until the 1930s, and now is the centrepiece of 'America's most inaccessible National Park' – yet every year 80,000 people make the journey by boat and sea plane to see it.

Sport fishing is a big attraction, especially in Islamorada, which is known for its spectacular sunrises. The 'backcountry' boats have raised platforms for shallow-water punting.

HAWAII

Latitude 21°18'N **Longitude** 157°47'W

Area 28,311 square kilometres (10,931 sq miles)

Status US state

Population 1,360,000

Capital Honolulu

Official language English, Hawaiian

Currency US dollar

The islands of Hawaii are extraordinary. They are parked out in the middle of the Pacific, some 3,000 kilometres (1,865 miles) from any continental landmass, and almost as far from any other islands. They rise like a row of buttons from a highly active faultline in the Pacific Plate, with seven inhabited islands in a string some 650 kilometres (405 miles) long – or 2,400 kilometres (1,500 miles) if all 137 islands are counted right to the far northwestern end of the chain. The biggest of the buttons, Hawaii Island itself, rises 4,205 metres (13,800 ft) at the summit of the extinct volcano, Mauna Kea, where some of the world's most powerful telescopes gaze deep into space. Somehow, about 1,700 years ago, Polynesian navigators managed to find these islands. Since 1959 they have constituted the 50th state of the USA, and they rank among the most prosperous of all the Pacific islands.

The culture of the islands remains distinctly Polynesian, albeit with plenty of American aspects enmeshed with it. *Leis* (garlands) of orchids or other tropical flowers are presented to honoured guests; mesmeric hula dances are performed by tattooed men in floral headdresses and women in grass skirts; steel guitars and ukuleles create the distinctive jangling sounds of traditional Hawaiian music. Of course, these have been adapted for the tourist industry, but the foundations are real.

The Kohala Coast on the northwest of the 'Big Island' of Hawaii is celebrated for its sunsets, beach resorts and golf courses.

TRAVELLER'S **TIPS**

Best time to go: The high season is the December–January holiday period, but this coincides with the wet season (October–March), which has more rainfall.

Look out for: Humpback whales congregate off Maui in November–May. Visitors can see them close at hand by joining an organized whale-watching tour.

Spouting Horn is an impressive blowhole in the Koloa district of Kauai. The sea has eroded away crevices in the lava crust, and the waves, surging in under the rock overhang, force water through the hole to create a spout of up to 15 metres (50 ft) high. The Spouting Horn features on the 16-kilometre (10-mile) Koloa Heritage Trail.

The tourist industry is huge. Some 6.5 million visitors come to Hawaii every year. Most of them go to Oahu, the site of the capital, Honolulu. There is much to see and do here: sunbathing and surfing on Honolulu's famous Waikiki Beach, and visiting some of the 21 Oahu state parks, or the Bishop Museum of Hawaiian history and natural history, or Pearl Harbor, scene of the notorious Japanese attack on the US Navy in December 1941. The Iolani Palace, built in the 1880s, is the only royal palace on US soil – home to the last monarchs of Hawaii, King Kalakaua and then Queen Liliuokalani, before she was ousted in 1893.

Oahu is in fact only the third biggest island. The largest, Hawaii itself, rests on a Pacific Plate hotspot, and there are two active volcanoes here, Mauna Loa and Kilauea, now encompassed by the Hawaii Volcanoes National Park: Kilauea is the smaller of the two, but has been constantly active since 1983, shooting out jets of steam, fountains of lava and rivers of molten rock that drip into the sea on the east coast. Over on the west coast, a monument marks the spot where Captain Cook met his end, in 1779, during an argument over the possession of a ship's boat.

Maui, the second largest island, lying between Oahu and Hawaii, and Kauai at the northwestern end of main islands, are quieter in mood, cherished for their beaches, dramatic volcanic landscapes and rich plant life. In Kauai, the wettest and greenest of the islands, vertical volcanic cliffs of the Na Pali Coast State Park are so coated with plant life that they look from a distance as if they are covered in felt. The rainforest scenes of *Jurassic Park* were filmed on Kauai.

The last in the chain of inhabited islands, 29 kilometres (18 miles) from Kauai, is Niihau, nicknamed 'the forbidden island'. Privately owned, it is home to some 200 mainly pure-blood Hawaiians, who try to maintain traditional Polynesian culture, speaking Hawaiian as their main language. Having no tropical flowers, they make *leis* from shells, as a source of income. They eschew all modern communications devices, such as mobile phones, radios and televisions. It is a kind of Utopian dream, lived out in the blue Pacific Ocean.

The most famous surfing areas – such as Waimea Bay and Sunset Beach – are on the North Shore of Oahu, where the waves challenge the world's top surfers. Quite who invented surfing is unclear, but Captain Cook saw it in Hawaii 1779, and earlier expeditions recorded seeing it in Tahiti.

TRAVELLER'S **TIPS**

Best time to go: In July–September the rainfall, humidity and temperature are lower and more agreeable. The heaviest rainfall comes in December–March.

Look out for: Wood carving: combs, bowls, masks, figurines of fish and birds, and the expressive *nguzunguzu* heads made as prow decorations for war canoes.

Dos and don'ts: Do remember this is a malaria zone: start protective medication before arrival. Also cone shells in the sea deliver a sting that can be lethal.

NEW GEORGIA

Latitude 8°15'S **Longitude** 157°30'E

Area 2,037 square kilometres (786 sq miles)

Status Island group of the Solomon Islands

Population 19,500

Main settlement Munda

Official language English

Currency Solomon Islands dollar

The Roviana Lagoon is peppered with dozens of tiny islands, beckoning visitors – who can paddle among them in kayaks or dugout canoes.

Picking up on a description long applied to the Solomon Islands, the national broadcaster – based in the capital Honiara, on the island of Guadalcanal – calls itself Radio Happy Isles. There are six main islands and island groups here, spread out across some 850 kilometres (530 miles), all lying about 1,000 kilometres (620 miles) east of New Guinea.

And of all the Happy Isles, the New Georgia Islands are celebrated as the most beautiful. Volcanic hills smothered in greenery frame huge, shallow lagoons. This is a paradise for divers and snorkellers. Coral reefs are thronged by myriad species of tropical fish; there are black coral gardens, forests of gorgonian sea fans, turtles, manta rays, giant clams. Uepi on the Marova lagoon in the south, the Roviana lagoon near

Munda in the centre and Gizo, on the island of Ghizo, are all famous diving centres.

Lodges and resthouses offer accommodation among the coconut palms, often in hideaway thatched bungalows, owned and operated by people who are noted for their friendly welcome and genuine hospitality. Visitors arrive by local plane services, but usually travel around by boat or canoe: they should come forearmed with a sense of adventure.

These islands were heavily fought over in the Second World War, and there are many mementoes of that time. Divers can visit the wrecks of aircraft, troopships, war tankers and landing barges, as well as the coral reefs that surround the islands. Plum Pudding Island, between the island of New Georgia and Kolombangara, is where John F. Kennedy was rescued by islanders in a dugout canoe in 1943, after his motor torpedo boat PT-109 had been rammed and sliced in two by a Japanese warship.

The inhabitants' fearsome reputation as headhunters kept colonial powers away from these islands, until Britain intervened in 1892. The Christian missionaries who then came here had more influence on steering the culture than any commercial exploitation, with the result that the islands remain remarkably unspoilt. And still today the Solomon Islands, and New Georgia in particular, are often whispered about as one of the South Pacific's best-kept secrets.

EASTER ISLAND

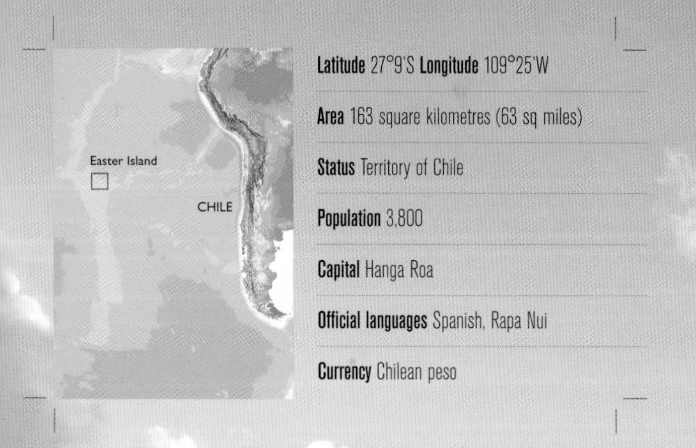

Latitude 27°9'S **Longitude** 109°25'W

Area 163 square kilometres (63 sq miles)

Status Territory of Chile

Population 3,800

Capital Hanga Roa

Official languages Spanish, Rapa Nui

Currency Chilean peso

'The Navel of the World' is the name that the early Polynesian inhabitants gave to their island, as well they might: their little lump of windswept volcanic rock is all that protrudes from this quarter of the southern Pacific Ocean. The nearest inhabited place is Pitcairn, 2,000 kilometres (1,250 miles) to the west. Chile, the country to which Easter Island now belongs, lies 3,700 kilometres (2,300 miles) to the east. This is about as remote as you can get, and it is a wonder that the first Polynesians ever found the place at all. At the tail end of the migrations that began in about 1500 BC, they probably arrived here in their large wooden sailing canoes some time between AD 700 and 1200, perhaps from the Marquesas Islands, 3,200 kilometres (2,000 miles) to the west.

The utter sequestration of Easter Island (known locally as Rapa Nui) is a fact that all modern-day visitors to Easter Island are aware of, as they fly over mile upon mile of empty ocean before this speck of land appears, bearing its broad strip of runway at Mataveri like a sticking plaster. Most people who come here are on a kind of pilgrimage. From their guesthouses and hotels they will spread out along the roads and tracks – by minibus, hired jeeps, on horseback, motor scooters and bicycles – to seek out the mysterious giant stone heads, totems of a rich and unique culture that rose and died in almost absolute isolation.

Some 70 *moai* statues dot the slopes of the volcano Rano Raraku, where the main quarry is located. Another 300 lie unfinished within the quarry.

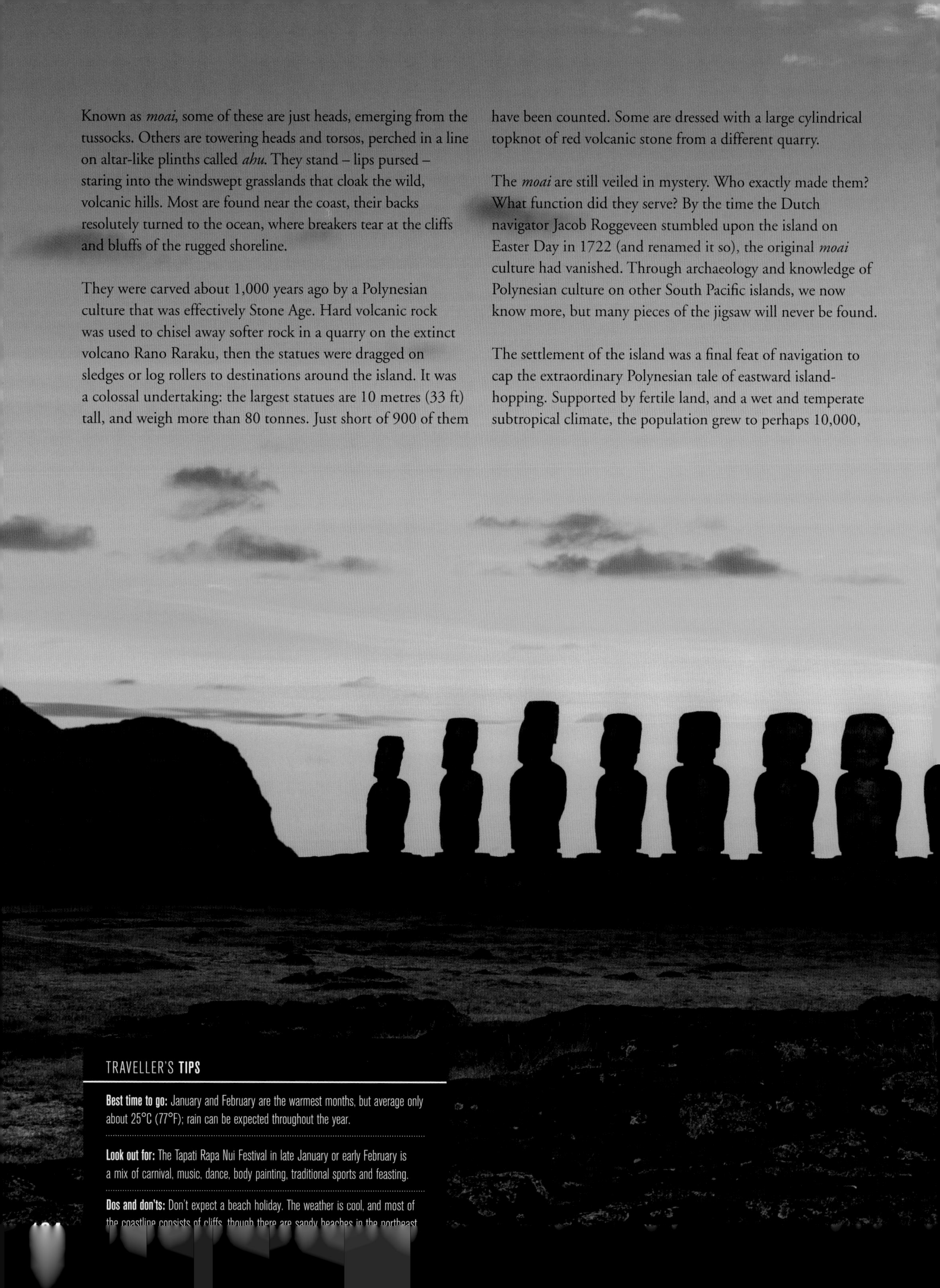

Known as *moai*, some of these are just heads, emerging from the tussocks. Others are towering heads and torsos, perched in a line on altar-like plinths called *ahu*. They stand – lips pursed – staring into the windswept grasslands that cloak the wild, volcanic hills. Most are found near the coast, their backs resolutely turned to the ocean, where breakers tear at the cliffs and bluffs of the rugged shoreline.

They were carved about 1,000 years ago by a Polynesian culture that was effectively Stone Age. Hard volcanic rock was used to chisel away softer rock in a quarry on the extinct volcano Rano Raraku, then the statues were dragged on sledges or log rollers to destinations around the island. It was a colossal undertaking: the largest statues are 10 metres (33 ft) tall, and weigh more than 80 tonnes. Just short of 900 of them have been counted. Some are dressed with a large cylindrical topknot of red volcanic stone from a different quarry.

The *moai* are still veiled in mystery. Who exactly made them? What function did they serve? By the time the Dutch navigator Jacob Roggeveen stumbled upon the island on Easter Day in 1722 (and renamed it so), the original *moai* culture had vanished. Through archaeology and knowledge of Polynesian culture on other South Pacific islands, we now know more, but many pieces of the jigsaw will never be found.

The settlement of the island was a final feat of navigation to cap the extraordinary Polynesian tale of eastward island-hopping. Supported by fertile land, and a wet and temperate subtropical climate, the population grew to perhaps 10,000,

TRAVELLER'S **TIPS**

Best time to go: January and February are the warmest months, but average only about 25°C (77°F); rain can be expected throughout the year.

Look out for: The Tapati Rapa Nui Festival in late January or early February is a mix of carnival, music, dance, body painting, traditional sports and feasting.

Dos and don'ts: Don't expect a beach holiday. The weather is cool, and most of the coastline consists of cliffs, though there are sandy beaches in the northeast even 17,000. Ancestral spirits protected the villages – represented by the *moai*. Clan rivalry may have driven islanders to create ever bigger statues: one colossally ambitious *moai*, 21.6 metres (71 ft) long, still lies unfinished in the quarry. Clan rivalry may also have induced them to knock the *moai* down. Captain James Cook noted this habit when he visited the island in 1774. By 1840 none was left standing; those standing today have been re-erected in modern times.

By the time of first European contact, the population of Easter Island had been reduced to about 4,000. Overpopulation had probably stretched resources to breaking point, resulting in deforestation, warfare, starvation and disease. The *moai* culture was replaced by the cult of the Bird Man, with a ritual race to collect eggs from the offshore islet called Moto Nui.

In the wake of European discovery, Easter Island took a dramatic turn for the worse, its people decimated by slaving raids, imported diseases and emigrations. By 1877 the population stood at just 111. In the late 20th century tourism revived the island's economy, and rekindled pride in its heritage. Now some 70,000 tourists come here each year, lured by the mystery of this strange and remote culture – and the secrets that its silent, pensive figures still refuse to yield.

Fifteen *moai* statues stand on a plinth at Ahu Tongariki, the island's most impressive line-up. Knocked over by a tsunami in 1960, they were restored in 1992–4 using a massive Japanese crane – a measure of the extraordinary achievement of the original builders.

GALAPAGOS

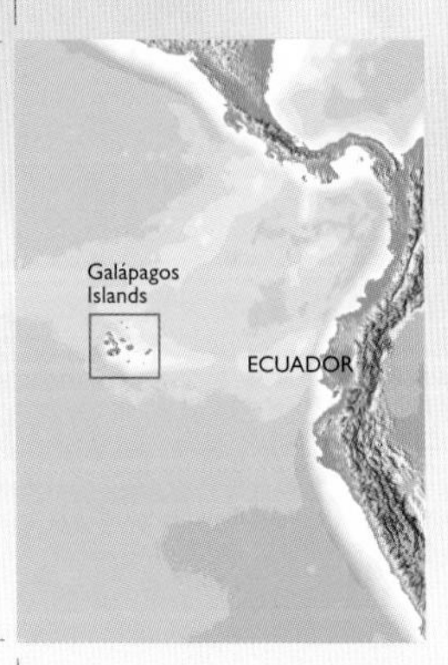

Latitude 0°40'S **Longitude** 90°33'W

Area 12,950 square kilometres (5,000 sq miles)

Status Province of Ecuador

Population 28,000

Capital Puerto Baquerizo Moreno

Official language Spanish

Currency US dollar

Sitting on the Equator in the Pacific Ocean 1,000 kilometres (600 miles) west of Ecuador is a scattering of small volcanic islands whose fame far outstrips their scale. And it is their very isolation that has brought this about. Because islands are isolated by water, they all have the potential to develop their own species of animals and plants. The Galápagos islands have achieved this to an exceptional degree. Here we find unique species such as marine iguanas, giant tortoises, flightless cormorants and the world's most northerly species of penguin.

In 1835, the 26-year-old British naturalist Charles Darwin spent five weeks on the islands, primarily as a geologist on the naval survey ship, HMS *Beagle*. Darwin's natural curiosity was aroused by the variations he noticed in individual species. He saw, for example that the 14 kinds of Galápagos finches had different beaks on different islands, and that this related to the kind of food that was available: small, strong beaks for crushing seeds; longer beaks for tearing fruit; thin and pointed beaks for picking insects out of their hiding places.

Darwin reasoned that animals with the best features to make use of the resources available would be the strongest: they were the ones that would survive hard times and reproduce – so those advantageous features would be passed to the next

TRAVELLER'S **TIPS**

Best time to go: The sea is calmest in December–May, but the animals are most active in June–August. September–November is least favoured.

Look out for: Go snorkelling or scuba-diving: the Galápagos islands are among the best in the world for diving, especially for seeing large marine animals.

Dos and don'ts: Many of the animals are remarkably docile but watch out for bull sea lions, which can be dangerously aggressive when protecting their harems.

Looking like dinosaurs, marine iguanas cling to the volcanic rocks of Fernandina Island, a landscape first thrust up from the sea floor 2–3 million years ago.

In its mating ritual, the male great frigatebird – here on Genovesa Island – inflates and contracts its red gular pouch with every breath. It is assumed that the size and colour of the pouch allows the female to gauge the health and vigour of the male, and his suitability for breeding. Frigatebirds range far and wide across tropical zones.

generation. This theory was later summarized as 'the survival of the fittest'. Winding the theory backwards, Darwin saw how all animals must have evolved over thousands of generations. They had not, after all, been placed on Earth readymade by God, but had developed through evolution. Darwin felt this theory might be so shocking that he hesitated for more than 20 years before he published *On the Origin of Species by Means of Natural Selection* in 1859.

There have been times in the past when the unique wildlife of the islands was distinctly threatened. Discovered by accident in 1535 by the Bishop of Panama, blown off course on his way to Peru, the islands were soon visited by whalers as well as pirates, who saw them as an open larder. They especially treasured the giant tortoises because they could be kept alive in

The sea has carved a natural arch out of the soft volcanic rock off Isla Darwin, the most northerly island of the archipelago.

the hold of ship and provide fresh meat for months: the name Galápagos means 'tortoises'.

In the 19th century, the islands were used as a penal colony. Now some 100,000 nature-loving tourists come here every year. This could in itself be considered a threat, but since 1959 the islands have been protected as a National Park, which now covers a full 97 per cent of the land area. Behaviour in the park is strictly regulated, and income from permits to enter the park contributes to conservation. Almost all the visitors travel around the islands on guided cruises, in boats of all shapes and sizes and budgets. The hundreds of islands and tiny islets each offer something different: a special mix of wildlife; nature walks; white-sand beaches that are perfect for swimming and snorkelling.

There are astonishing views from the volcanic peaks. However, the highest of all, Wolf Volcano at 1,707 metres (5,600 ft), on the largest island, Isabela, is difficult to climb because it is still active, and the recent eruptions have left the sides very steep and crumbly – a reminder of the raw, elemental forces that continue to shape these islands.

It is not just the unique and extraordinary animals of the Galápagos that the visitors come to see. The islands are rich in wildlife of all kinds: pelicans, booby birds, albatrosses, seals, sea lions, turtles, bright orange Sally Lightfoot crabs, the luminescent plankton that envelope night swimmers like stardust. But somehow the extraordinary quality of some makes no creature seem ordinary, however familiar. That is perhaps the greatest magic of the Galápagos islands.

TAHITI

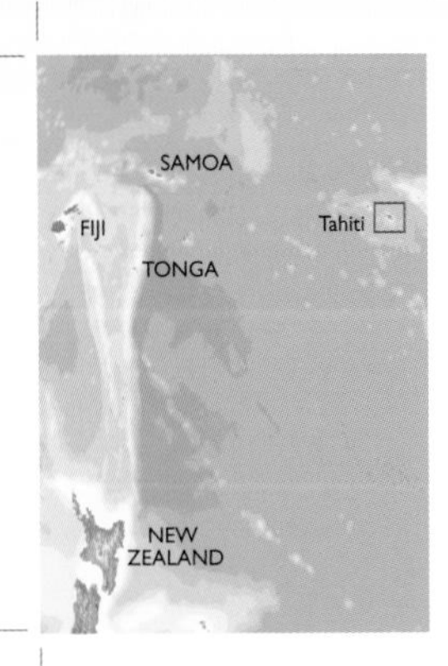

Latitude 17°38'S **Longitude** 149°27'W

Area 1,045 square kilometres (403 sq miles)

Status Island of French Polynesia

Population 178,000

Capital Papeete

Official language French

Currency CFP franc (Comptoirs Français du Pacifique)

For many people, Tahiti is the ultimate travel dream. They picture palm-shaded beaches; clear, clean, bath-warm lagoons edging towards a deep blue horizon; an Eden of lush vegetation; people dressed in brightly printed sarong-like *pareos*; women of unlikely beauty with a flower tucked behind the ear. And they are surprised to find – from the moment that they arrive at the airport and are handed *tiare* (gardenia) flowers to the rhythm of a ukulele band – that, incredibly, this is all true. Tahiti's very remoteness adds to its allure. It is the furthest east of the major island groups of the South Pacific,

TRAVELLER'S **TIPS**

Best time to go: The weather is cooler and drier in May–October. June–August is peak season. It is hotter and more humid - but less busy - in November–April.

Look out for: Monoi oil, the extract of local gardenia (*tiare*) flowers mixed with semi-hard coconut oil. Used as a skin moisturiser, it has an beautiful scent.

Dos and don'ts: Don't be in a rush. The Tahitian attitude to life can be summed up as *aita pea pea* ('not to worry').

Tahiti is the largest island of French Polynesia, and serves as the centrepoint for 118 scattered islands, many of them offering tropical perfection – like this beach on the island of Rangiroa.

halfway between New Zealand and North America. Further east lie only tiny Pitcairn and Easter Island. Land here is just a scattering of specks in the vastness of the oceans.

Tahiti itself, however, is a relatively large speck. Shaped like a fat beaver swimming towards the smaller island of Moorea, the body is Tahiti Nui and the wide, tail-like peninsula Tahiti Iti. Tahiti Nui is a straightforward, classic, cone-shaped volcanic island, rising to the double peak of its old, long-extinct craters, with Orohena the highest at 2,241 metres (7,352 ft). One of the great pleasures of the island is to follow a route up to the peak, through woods of astonishingly rich profusion, with every shade of glossy green, and every leaf-shape conceivable, stooping off at cascades tumbling into jungle-shaded pools.

Tahiti is not just a modern fantasy: it seems to conform to some hardwired dream of what paradise looks and feels like. The very first European visitors were struck by this. Their ships were quickly surrounded by canoes, and a people so welcoming and accommodating it was hard not to take advantage. Compared with the harsh conditions of discipline, deprivation and sheer hard physical labour on board large sailing vessels, the life of the islanders seemed idyllic. Many crewmen jumped ship.

Captains were faced with a cruel dilemma: how to maintain discipline, and not cause offence to their hosts. The great French explorer Louis-Antoine de Bougainville (whose name is remembered in the beautifully coloured tropical plant bougainvillea) came here in 1768, and wrote eloquently of the impact of the welcome by beautiful, near-naked women possessing few of the usual Western social inhibitions. 'I put it to you: how, in the midst of such a show, do you keep 400 young French sailors, who have had no sight of a woman for six months, focused on their work?' Bougainville called the island New Cythera, after the island of Aphrodite, the Greek goddess of love. This too was the cause of the mutiny on HMS *Bounty*, which stayed in Tahiti for five months in 1788–9, collecting breadfruit plants to take to the Caribbean islands. Many of the crew struck up relationships with Tahitian women; Lieutenant Fletcher Christian married one. The contrast between naval life and the idyll they experienced on Tahiti was just too great to bear, with tragic consequences of lonely exile on Pitcairn.

In fact the idyll collapsed for the Tahitians too, through Western contact. In 1797, the London Missionary Society stepped in to try to stem the degradation, and a dynasty of Tahitian kings called Pomare brought some stability, before Pomare IV was persuaded to allow the island to become a

French Protectorate in the 1840s. The dream never went away, however. It drew the French painter Paul Gauguin here in the 1890s, before he ended his days in the Marquesas Islands to the northeast. It draws many thousands of visitors still. Few are disappointed.

Many hotels on the Tahitian islands offer overwater bungalows, where guests can sleep to the sound of the waves lapping beneath them, and walk down a ladder into the sea. The traditional thatch of pandanus leaves helps to keep the interior cool.

Founded in 1998, Les Grands Ballets de Tahiti brings a contemporary twist to traditional dances. Using trademark features of Polynesian dance, such as mesmeric hip-shaking of the women and warrior-aggression of the men, the troupe presents a vigorous stage show featuring forty-five dancers, twenty musicians and six singers.

BORA BORA

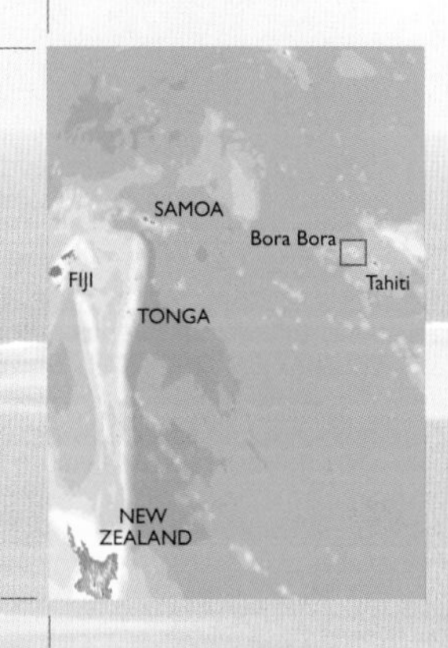

Latitude 16°29'S **Longitude** 151°44'W

Area 29.3 square kilometres (11.31 sq miles)

Status Island of French Polynesia

Population 8,900

Capital Vaitape

Official language French

Currency CFP franc (Comptoirs Français du Pacifique)

If you dream of a dream island in the South Pacific, it probably looks like this. A broken ring of low palm-crested coral islands and reefs encircles a lagoon of every translucent shade of blue and turquoise imaginable. In the middle stands a volcanic cone with dense, ruffled vegetation rising all the way to its dramatically shattered twin peaks. The beaches are coral-white, the water crystal-clear, temperature constant and warm, softened by easterly trade winds.

'Pearl of the Pacific' is the sobriquet given to Bora Bora, and its beauty now draws 100,000 visitors a year, many on honeymoon, from the USA, Europe, Australia and Japan. They mostly fly in via Papeete on Tahiti, which lies 240 kilometres (150 miles) to the southeast. They come for the beach life, to take boats out into the lagoon, and to snorkel and scuba-dive among the coral reefs; in the evening they are entertained by displays of sensual Polynesian dance.

Bora Bora is a textbook case of coral-island construction. About 3 million years ago, a volcano thrust up from the ocean floor, creating the peaks of the central island, plus the two smaller islands of Toopua and Toopuaiti. Living coral then colonized the shallows around its shores. Very slowly the central peak, with its mighty plugs of rock, began the natural process of sinking back towards the ocean floor, but as it did

TRAVELLER'S **TIPS**

Best time to go: All year. The weather is cooler and drier in May–October, hotter and more humid in November–April. June–August is the busiest time.

Look out for: Outrigger canoes take visitors on lagoon trips to see the coral gardens, manta rays, giant clams, and reef-shark feeding-time.

Dos and don'ts: Do hire a bicycle and take the 30-kilometre (19-mile) ring road around the main island.

The white sail of a yacht underlines the extraordinary intensity of the cobalt blue in a deep-water channel in Bora Bora's lagoon, off Motu Pitiaau, on the southern tip of the outer reef.

so, the coral built ever higher to remain in the warmth and sunlight of the shallows. Thus a ring of coral formed, separated from the sinking peak by a shallow lagoon. Over time, some of the reefs and shallows became silted up, creating islands – called *motus* in Polynesia – with enough soil to host shrubs and palms.

Polynesians settled on the island about 1,500 years ago. They called it Pora Pora, meaning firstborn, referring to their tradition and mythology that made Raiatea, 43 kilometres (27 miles) to the north, the parental island. More than 40 outdoor temples, called *marae*, with stone altars, can still be seen on Bora Bora. The Tahitian language does not use the b sound, but 'Bora Bora' is what the first European explorers – such as the Dutchman Jacob Roggeveen in 1722 and James Cook in 1777 – thought they heard. It doesn't use the sound l either. When the British sailor James O'Connor survived the wreck of his whaling ship, the *Matilda*, in 1792, and settled with a Polynesian wife in the south of the main island, the place was named after the ship. Pointe Matira is now renowned as the location of the best beaches on the island.

The French annexed Bora Bora in 1895, as they extended their control over French Polynesia. The sleepy beauty of this outpost came in for a rude shock during the Second World War when US forces turned Bora Bora into a refuelling and

supply station, built a runway on Motu Mute island (the site of the modern airport), filled the lagoon with troopships and stationed some 4,400 personnel here. They also placed eight enormous naval guns on the main island in readiness for a Japanese attack that never came. The guns are still there, preserved as a tourist attraction.

The first ever overwater bungalows in the world were built at Hotel Bora Bora, which opened in 1961, and this is now the dominant style of accommodation on the main island and on the *motus.* With views out over the blue lagoon and to the volcano, this is the perfect place to witness how a dream island can in fact become reality.

Bora Bora has pretty, pastel-painted churches in colonial style, like this one in Vaitape. As in the rest of French Polynesia, the London Missionary Society influenced religion in Bora Bora, arriving here in 1820 and establishing the Protestant church as the main religion, later reorganized under the Eglise Evangélique de Polynésie Française.

Bora Bora pioneered the trend for overwater bungalows back in the 1960s. Behind rise the naked peaks of Mount Pahia and Mount Otemanu, a volcanic plug which, at 727 metres (2,385 ft), marks the island's highest point.

TRAVELLER'S **TIPS**

Best time to go: The warmest months are September–March, but December–March is the rainy (and cyclone) season. June–August can be cool.

Look out for: The Tjibaou Cultural Centre in Nouméa is a striking piece of prestige architecture by Renzo Piano: its collection provides an insight into Kanak culture.

Dos and don'ts: Do try *bougna*, a dish of meat or fish with fruit and vegetables, parcelled in banana leaves and slow-cooked on hot rocks beneath a pile of earth.

NEW CALEDONIA

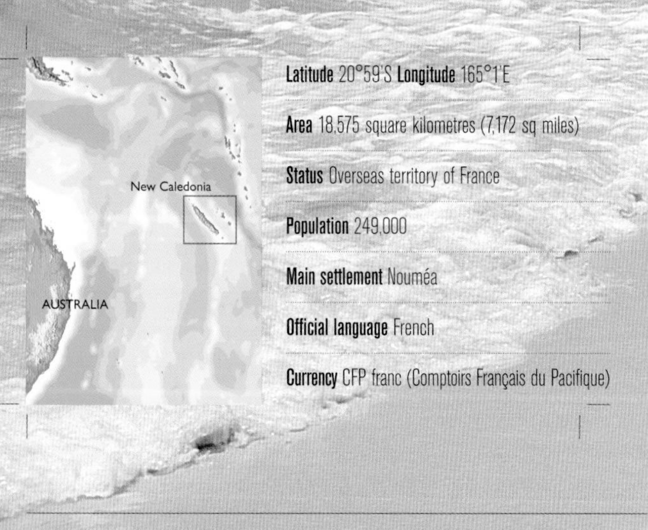

Latitude 20°59'S **Longitude** 165°1'E

Area 18,575 square kilometres (7,172 sq miles)

Status Overseas territory of France

Population 249,000

Main settlement Nouméa

Official language French

Currency CFP franc (Comptoirs Français du Pacifique)

Ouvéa, one of the Loyalty Islands, has an unbroken white-sand beach stretching 25 kilometres (15 miles). It borders a wide and shallow lagoon, which turns the sea to a scintillating turquoise.

In the *pilou* dance, muscular Kanak men – dressed in trailing grass skirts, feathers, shell beads and body paint – stomp, swing and rustle in rhythmic formation to the beat of percussion, chants, trills and whistles. They reach an almost trance-like state. Well performed, the *pilou* has an elemental power.

There is something elemental about the whole of New Caledonia. The mountainous landscape of the main island, Grande Terre, forms a long slug of land, 400 kilometres (250 miles) long and 50 kilometres (30 miles) wide, folded and lifted by the collision of the Pacific and Indo-Australian plates. Much of the interior is covered in arid scrub, but here and there rivers cut through, tumbling to the sea in chain of waterfalls and pools. On the coast – notably near the town of Hienghène – the sedimentary rock has been eroded into savage, towering shapes, the sacred homes of ancestor spirits. The island is encircled by the reefs – second in size only to

Australia's Great Barrier Reef – and the world's largest lagoon. Whales play off the coast in August and September, having journeyed up from the Antarctic.

Traditional outrigger canoes, or *pirogues*, with triangular sails, take local fishermen around the reefs, or to the scattering of coral islands to the north and south, or even to the Loyalty Islands – Ouvéa, Lifou and Maré – that lie in a row some 100 kilometres (60 miles) to the northeast of Grande Terre. Here, many people still live in round, thatched houses, entered through a low, carved doorway.

The Kanaks, the original inhabitants of Melanesia, have been exposed to Western culture since Christian missionaries arrived in the 1830s: the word *kanak* is actually Hawaiian, meaning 'man', and was introduced by Europeans. The islands were then settled, farmed and mined for nickel by the French.

La Nouvelle Calédonie remains a French territory. But it was James Cook, the British explorer, who named the islands New Caledonia, because the mountainous landscape near Balade on the northeast coast, where he anchored in 1774, reminded him of Scotland (for which Caledonia was the Roman and poetic name). He also named the Isle of Pines in the far south, noting that its tall pines – the rare candelabra pines (*Araucaria columnaris*) – might be used for replacement masts. The long isolation of New Caledonia has given it some 2,500 plants found only here. Many of these can be seen in the Parc Provincial de la Rivière Bleue near Yaté, in the southwest of Grande Terre.

Today New Caledonia is cherished by hikers, sailors and divers, and those in search of a getaway experience. The Isle of Pines is the jewel in the crown, earning the nickname *l'île la plus proche du paradis* ('the closest island to Paradise').

MALDIVES

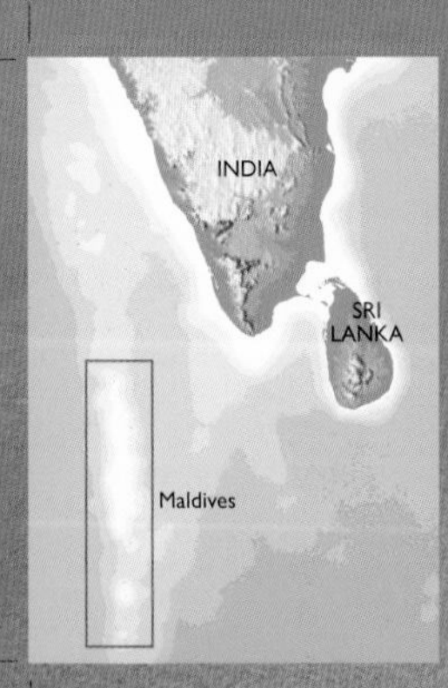

Latitude 3°12'N **Longitude** 73°13'E

Land area 298 square kilometres (115 sq miles)

Status Independent island republic

Population 396,500

Capital Malé

Official languages Dhivehi

Currency Rufiyaa (US dollars in resorts)

Sun, sea and… honeymoon. The Maldives offer the ultimate in dream islands, if that dream is a peaceful coral atoll, white-sand beaches and palm trees beneath a blue sky dabbed with cottonwool clouds.

Since the 1970s, the Maldives have concentrated on delivering that dream, creating nearly 100 upmarket resorts with individual bungalows or villas built out over the idyllic atoll lagoons. The Maldives have the space, and they have the setting. There are 1,192 islands in this archipelago, divided

Sunset brings a change of pace to a set of luxury overwater villas on the northern atoll of Maalhosmadulu Uthuruburi – more often referred as Raa, under the Dhivehi alphabetic coding used to simplify the atoll names.

into 26 atolls, stretching 820 kilometres (500 miles) north to south. Most of the islands are tiny, and add up to a minuscule combined land area compared with the 90,000 square kilometres (34,750 sq miles) of territorial waters. Only 220 of the islands are inhabited by Maldivians. It is the uninhabited islands where tourist development has taken place.

The atolls sit upon the old peaks of a volcanic mountain chain thrust up millions of years ago along a faultline on the western edge of the Indo-Australian Plate. That faultline has moved

TRAVELLER'S **TIPS**

Best time to go: January to April: this is also the best time for divers. Rainfall is higher in the monsoon season from April to October, particularly June–August.

Look out for: Dhoni boats: these are the traditional square-rigged sailing boats, which have a shallow draught especially suited to entering lagoons.

Dos and don'ts: Note that it is forbidden to bring alcohol into the country (it is available at the resorts) or to take any seashells or coral home.

westward, so the islands are no longer volcanic. Instead, the peaks have dropped back into the sea, leaving just circlets of coral reefs – classic coral atolls.

The highest point in the Maldives is just 2.3 metres (7 ft 7 in) above sea level – making it the lowest-lying country in the world. It is easy to see how the South Asian tsunami of 2004 – with waves up to 4.3 metres (14 ft) high– caused widespread devastation here. There are also long-term concerns that the rising sea levels associated with global warming will eventually overwhelm the Maldives.

For visitors to the luxury resorts, there may be little evidence of trouble in paradise. As a matter of government policy, visitors are kept largely separate from the local population, apart from hotel staff, unless they specifically choose to go to the capital, Malé. This is where a quarter of Maldivians live, and is the only sizable town on the archipelago: a grid of medium and high-rise streets crammed onto its island in the middle of the archipelago. Most tourists bypass it, travelling directly from the airport to their resort, by boat or light aircraft, and as a result have little contact with the Sunni Muslim population.

The culture of the Maldives has developed over 2,500 years, from the days of the first settlers who came from southern India and Sri Lanka, 700 kilometres (435 miles) to the east. The scattering of islands formed part of the Buddhist and Hindu empires of the Indian subcontinent, but they were also on the route between the Arabian Gulf and Sri Lanka, and valued in their own right as a source of cowrie shells (used as currency), dried tuna fish, and coir – coconut fibre used to

make ropes. This link brought Islam, and from 1153 to 1963 the Maldives were ruled as a Muslim Sultanate, becoming a British Protectorate in 1887. Full independence was granted in 1965, when tourism had barely been considered an option.

These circumstances have permitted the Maldivians to be the masters of their tourist industry, which now attracts some 600,000 visitors a year – nearly twice the total population of the islands. They have focused on luxury resorts, appealing to those travellers prepared to spend small fortunes for the comforts of five-star accommodation, the lapping of the blue seas, the talcum softness of white sand between their toes, the timeless world of the coral reefs – the ultimate island dream.

The coral reefs are a paradise for scuba-divers, and many resorts are geared up to cater for them. There are 2,000 species of fish here, with expressive names such as pufferfish, sweetlips, angelfish and scorpionfish. Divers may also encounter whale sharks, the gentle filter-feeding giants that can reach 12 metres (40 ft) in length.

Looking eastward over Meeru Island in the Maldives. Coral sands provide a foothold for coconut palms. The palms create a local ecology that over time gathers a shock of greenery in otherwise hostile conditions.

SEYCHELLES

Erosion has created smooth, curtain-like shapes in the granite outcrops on Anse Source d'Argent, on the island of La Digue, adding a sculptural feature to the white-sand beaches.

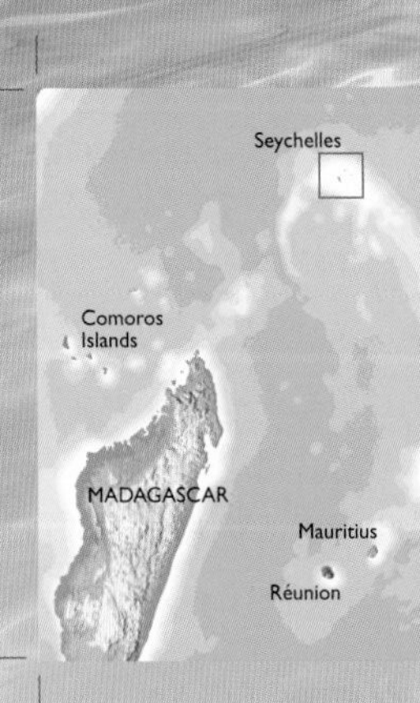

Latitude 4°37'S **Longitude** 55°27'E

Land area 451 square kilometres (197 sq miles)

Status Independent island republic

Population 84,000

Capital Victoria

Official languages French, English, Seychellois Creole

Currency Seychellois rupee

There are exclusive islands in the Seychelles – Frégate, North Island, Cousine, Desroches – where everything possible is done to create the ultimate tropical-island dream. Private, hideaway villas built in native mahogany, exquisitely furnished, open out onto views of soft white-sand beaches fringed by palms, coral reefs and warm, crystal-clear, turquoise seas. They have private infinity pools, decking belvederes, outdoor bathrooms under the stars, impeccable service and food of Michelin-star quality. They are also surrounded by mini Gardens of Eden. The preservation of the ecology is a

TRAVELLER'S **TIPS**

Best time to go: Any time of year is good: temperatures hover constantly around 27°C (80°F). Mid-November to mid-February is the rainy season.

Look out for: Woodland nature walks: there are marked paths through the Morne Seychellois National Park, Mahé, and the Vallée de Mai Nature Reserve, Praslin.

Dos and don'ts: Do watch your possessions. In recent years the Seychelles have suffered a drugs problem that has led to a rise in crime, particularly on Mahé.

high priority. The Seychelles government and the owners of private islands do everything they can to protect and promote their unique species – the Seychelles magpie robin, warbler, paradise flycatcher, black parrot, coco de mer palms, the Aldabra giant tortoise.

There are 115 islands in the Seychelles, sprinkled liberally in the Indian Ocean to the northeast of Madagascar and just south of the Equator. At the heart of them are the granitic Inner Islands, which include the two largest, Mahé and Praslin, and their 40 satellites such as La Digue, North Island, Frégate, Cousin and Cousine. Six tiny islands dot the Ste-Anne Marine Park, home to 1,000 species of tropical fish. The granite hearts of the Inner Islands give them elevation, often clothed in woods and shrubs, but the rims are still frosted with beaches of white sand. Mahé has the small capital of Victoria, and the international airport – and also its fair share of luxurious resort hotels. Further out, to the southwest, are coral islands, such as the Amirante Group (which includes Desroches), the Farquhar Group, and – 1,100 kilometres (700 miles) from Mahé – the Aldabra Islands. None rises more than 5 metres (16 ft above sea level). Few are inhabited. Only two of the islands, Mahé and Praslin, have cars. On La Digue, transport is by ox-cart or bicycle. Elsewhere, the islands are so small that cycling and walking are all that are needed.

Aldabra has been set aside as a nature reserve, protecting above all its population of Aldabra giant tortoises and nesting turtles; visitors need a special permit to land. They can, on the other hand, visit Bird Island, in the far north of the Seychelles

The Aldabra giant tortoise is one of two surviving species of giant tortoise, the other being from the Galápagos. Its stocky legs support a heavy shell up to 120 centimetres (4 ft) long. The tortoises were hunted for food by passing ships until, by 1900, they were on the verge of extinction; now there are 150,000 on the Aldabra atoll alone.

group. As the name suggests, it is a bird sanctuary, and here thousands of sooty terns, fairy terns, frigatebirds and common noddies gather, and hawksbill turtles and green turtles come to nest.

For centuries the Seychelles were uninhabited, visited only by passing Arab traders on their way to the East African coast; Kenya lies 1,600 kilometres (1,000 miles) to the west. The Portuguese navigator Vasco da Gama spotted the Amirantes Group in 1502 and named them *Ilhas do Almirante*, Admiral Islands. The French claimed the group in the 18th century, and when the French East India Company was granted formal possession in 1776, it named the islands after Louis XV's controller-general of finances, Jean Moreau de Séchelles.

The British and French squabbled over the islands during the Napoleonic Wars, and they fell to Britain (along with Mauritius) in the Treaty of Paris of 1814. Nonetheless, the local language of the islands has remained the French-based Seychellois Creole (or Seselwa). The islands survived on a plantation economy, producing cinnamon, vanilla and copra (coconut meat used to make oil), importing labour mainly from Africa, but also from India and China. Then, in 1976, they were granted independence – just as the Seychelles' new personality as the world's dream islands began to unfurl.

Endemic palms dominate the ecology of the Vallée de Mai nature reserve on Praslin, the second-largest island in the Seychelles. The seeds produced by one particular species of endemic palms, the coco-de-mer, are the largest in the animal kingdom.

MASCARENE ISLANDS

Latitude Réunion: 21°6'S Mauritius: 20°12'S
Longitude Réunion: 55°31'E Mauritius: 57°30'E

Area Réunion: 2,500 square kilometres (965 sq miles)
Mauritius: 2,040 square kilometres (787 sq miles)

Status Réunion: Overseas department of France
Mauritius: Independent republic

Capital Réunion: St-Denis; Mauritius: Port Louis

Currency Réunion: Euro; Mauritius: Mauritian rupee

Mauritius and Réunion are like twins. They sit in the Indian Ocean to the north of the Tropic of Capricorn, just 225 kilometres (140 miles) apart, and about 800 kilometres (500 miles) east of Madagascar. They are roughly the same size, Réunion slightly the larger. Both belong to the volcanic archipelago called the Mascarene Islands – and both are dream holiday islands. Mauritius is a relatively low-lying island, reaching its highest point at 828 metres (2,717 ft), at the Piton de la Petite Rivière Noire. Fringed with glorious white-sand beaches, it has developed a reputation for deliciously luxurious resort hotels: private villas with infinity pools, spa treatments and butler service. When travellers dream of Mauritius, they dream of pampering.

Réunion has a much more rugged and steeper profile, rising to 3,069 metres (10,069 ft) at Piton des Neiges (Snow Peak, for indeed it does occasionally snow on the summit). It also has the most active volcano of the islands: Piton de la Fournaise (Furnace Peak). Visitors come to Réunion not so much for luxury as activity holidays – scuba-diving, surfing, paragliding and, above all, hiking, especially the tours of the three Cirques – Cilaos, Salazie and Mafate – formed by old volcanic craters, high in the centre of the island.

Réunion and Mauritius shared a fair bit of their early history. As uninhabited islands, they were visited by Arab traders, and then by Portuguese navigators seeking a route to India in the

While some of the Mascarene Islands boast unrivalled beaches, the appeal of Ile de Réunion is its breathtaking landscapes – such as the Grand Galet waterfall, pictured here. Réunion is also home to a still-active volcano, Piton de la Fournaise.

early 16th century. The Dutch settled on Mauritius in 1638, having named the island after Prince Maurits of Nassau, the Dutch head of state. Hunting and the introduction of new animals decimated the native species, and brought about the extinction of the giant flightless pigeon, the dodo.

The French took over both islands later that century, establishing the French place names and the French-based creole languages still found in both islands. Then in 1810, during the Napoleonic Wars, Britain seized control of both, but gave Réunion back to France in 1815. In the post-slavery era, the plantation economy was sustained by indentured labour brought in from India, Malaya and China, which explains the ethnic mix of the populations today. Réunion is still French. In fact it is a part of France: since 1946 it has been an overseas *département*, like Martinique and Guadeloupe in the Caribbean. Mauritius, on the other hand, has been an independent republic since 1968.

As for the Mascarene Islands, they encompass a long and scattered chain that sweeps down to the east of Madagascar from the Seychelles. They are all islands that have stood on a shifting volcanic hotspot: over a period of 35 million years, one by one they have risen up from the ocean floor as volcanoes, then sunk back as the hotspot has moved on. The hotspot is currently beneath Réunion, under the Piton de la Fournaise.

The Mascarene Islands were named by the Portuguese navigator Diogo Rodrigues after the explorer Pedro Mascarenhas, who became Captain-Major of Malacca and Viceroy of Goa in the early days of Portuguese expansion.

The island group not only includes Mauritius and Réunion, but also Rodrigues, named after Diogo: just 109 square kilometres (42 sq miles), and with a population of around 40,000, it is a dependency of Mauritius, and lies 560 kilometres (348 miles) to its east. With a large lagoon attached to the volcanic island, it specializes in low-key 'green tourism', and diving and activity holidays. The Mascarene Islands also include the Cargados Carajos Shoals, a set of about 16 small islands, islets, sandbanks and lagoons to the northeast of Mauritius. They are only occasionally inhabited by passing fishermen – and offer the really intrepid travellers the ultimate desert island dream.

TRAVELLER'S **TIPS**

Best time to go: Mauritius and Réunion are warm all year round, but it is hotter and drier in May–November. The rest of the year is wetter and more humid.

Look out for: Indian influences in the food: curry (*kari*), spicy *rougail* (fish or meat stew), samosas, roti (flatbread). Mauritius and Réunion produce rum.

Dos and don'ts: Avoid mosquito bites, which have been responsible for dengue fever and flu-like chikungunya and (in Mauritius) malaria.

A tree on Roches Noires, on the northeast coast of Mauritius, graphically shows the direction of the prevailing winds. Here white sand contrasts vividly with the black volcanic rocks that have earned the beach its name.

MADAGASCAR

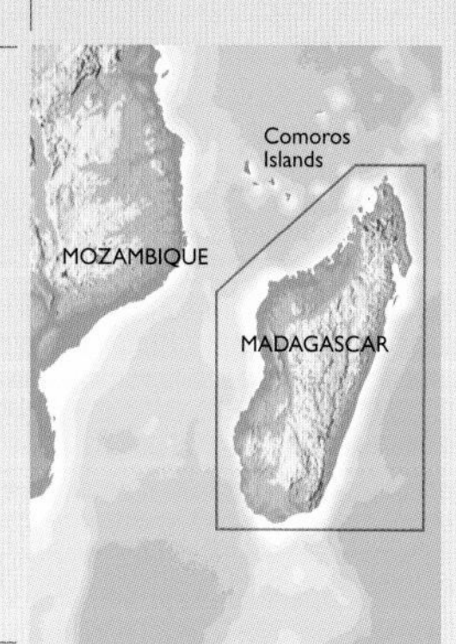

Latitude 18°55'S **Longitude** 47°31'E

Area 587,041 square kilometres (226,656 sq miles)

Status Independent republic

Population 20,653,000

Capital Antananarivo

Official language Malagasy, French

Currency Malagasy ariary

Madagascar lies just 400 kilometres (250 miles) off the coast of Africa, and yet it is a world apart. The Madagascans, for the most part, owe their ancestry not to the continent beside them, but to Indonesian islands 5,000 kilometres (3,000 miles) to their east. Quite how this happened remains a mystery, but it seems that about 2,000 years ago, migrants made this journey, and came to settle in Madagascar.

They arrived on an island that was already substantially different. Separated from Africa for 150 million years, it had developed its own set of animals and plants – and this remains one of the island's greatest attractions today.

About 80 per cent of Madagascar's animal and plant species are found nowhere else in the world. This includes, most famously, the lemurs – primates that are only distantly related to monkeys. There are about 100 different species and subspecies of lemur, from the elegant and acrobatic ringtailed lemur, to the Madame Berthe's mouse lemur, which at about the size of a human fist, is the world's smallest primate. Most

A dirt track road near Morondava, on the east coast, is turned into a grand and surreal avenue by rows of Grandidier's baobab trees – the most magnificent of Madagascar's six species of baobab.

TRAVELLER'S **TIPS**

Best time to go: November–April is hot and rainy, and May–October is cooler and drier, but the effects are localized and also depend on altitude.

Look out for: Vanilla pods are very cheap by Western standards, and of excellent quality; they are often sold in markets.

Dos and don'ts: Do be aware that crime is a growing problem, from pickpockets in Antananarivo to carjacking and nighttime banditry on country roads.

Madagascar has about half of the world's species of chameleon: this is a panther chameleon, found in the Lokobe Reserve on Nosy Be island. Chameleons came under threat in Madagascar because of the belief that their independently rotating eyes could look into the past and into the future, giving them supernatural powers.

of them are most active at night, which is why they were given their name: *lemures* is Latin for 'ghosts'.

Other animals unique to Madagascar include the narrow-striped mongoose and the tenrecs, a diverse range of animals resembling shrews, mice and hedgehogs. There are no large predators in Madagascar, but there is the fossa, a pugnacious, muscular and short-limbed relative of the mongoose and civet cat that prowls about like a panther sawn off at the knees. Groups of these animals have been known to attack cattle, and even humans.

Madagascar's plant life is as individual as its fauna. Of the eight species of baobab tree in the world, six of them are found in Madagascar (Africa has one, and Australia has the other). The trunks of these trees act like giant water reserves, and when they die they transform into a heap of debris rather than timber. Madagascar has 960 species of orchid, 75 per cent of which are found only on the island.

Ringtailed lemurs are just one of several species of lemur that are protected and observed for research at the privately owned Berenty Reserve, in the far south of Madagascar – one of the country's most popular and most visited nature parks.

Madagascar is a huge country, the fourth largest island in the world. The east coast is generally green and forested, with a humid climate. Most people live on the cooler central plateau, which includes the hilly capital, Antananarivo. Farming is the main pursuit: the staple is rice, which grows in emerald-green paddy fields, prepared by ploughs drawn by buffalos. Rain falls on this central plateau, creating a rainshadow to the east and south of the country, so this consists mainly of dry scrubland, savannah or desert. But pockets of shade and moisture can create oases, as in the little canyons, full of greenery and life, in the otherwise arid Isalo National Park.

The French ruled Madagascar from 1896 until the island gained its independence in 1960. They developed mining and export crops, such as coffee, sugar, cloves and vanilla. Their legacy of infrastructure has not been maintained, which means that travelling around Madagascar is not easy. There are a few railway lines on the east coast, connecting Antananarivo. Most of the roads are rudimentary.

Many visitors therefore choose to join guided tours, which take them to the national parks, and connect with the beach resorts and diving centres, such as the islands of Nosy Be in the northwest and the Ile Ste-Marie in the east – a noted spot for watching the migrations of humpback whales. Most of the travellers who come here are drawn to Madagascar because of the dream of encountering its unique wildlife, and most come away enchanted.

PHUKET

Latitude 7°53'N **Longitude** 98°23'E

Area 543 square kilometres (210 sq miles)

Status Province of Thailand

Population 348,500

Capital Phuket City

Official language Thai

Currency Baht

From beneath the shade of his driftwood shelter, the fisherman is fanning the embers of a fire. The sweet, acrid scent of roasting crustacean shell drifts across your vision of the warm sea where you have just spent the morning bathing, snorkelling and lying on the soft white sand. A light breeze rattles the fronds of the palm trees, and every now and then makes them raise their heads and sigh. At your feet, the pink trumpets of morning glories nod from their vines as they venture out onto the beach from the turf border of the shore. Lunch will be served on a picnic bench with slices of fresh lime, and a dish of fiery chilli sauce.

TRAVELLER'S **TIPS**

Best time to go: November–February is the dry ('cool') season, and the most pleasant months. The monsoon season (wet, hot and sticky) is May–October.

Look out for: Sunset at the Giant Buddha, for magnificent views. Wear 'respectful' clothes: no beachwear or T-shirts, a sarong rather than a short skirt.

Dos and don'ts: Do be vigilant when swimming in the sea: the currents can be vicious, especially in the monsoon season, and there have been fatalities.

This island fantasy is available at Phuket, Thailand's largest island. Unlike most of the country, which hooks around the Gulf of Thailand with shores looking south and east, Phuket is way down south on the Malay Peninsula and looks out west into the Andaman Sea, in the direction of India. Almost all the best beaches on the island also face in that direction – and the far southwestern promontory, between Nai Harn Beach and Rawai Beach, is celebrated for its dramatic pink-and-orange sunsets. Phuket has idyllic corners, but this is no desert island; it is linked to the mainland by road bridges and has an international airport. Slap bang in the middle of the west coast is Patong, the main tourist centre – busy, brash and raucous. For those who want high-action holiday fun, Patong and Phuket's many beach resorts are geared up to deliver, offering windsurfing, kite surfing, scuba-diving, golf, discos and night clubs. There is also a broad range of visitor attractions when a surfeit of beach life threatens: an aquarium, a crocodile and alligator farm, a butterfly farm. A recent addition is the Giant Buddha, a massive marble-clad statue that rises from a hilltop in the rainforest close to the island's inland capital, also called

A traditional boat stands ready for service – to go fishing in the shallow seas, or to take visitors to one of the many small islands that lie to the south and east of Phuket.

Phuket. The Buddha is 45 metres (148 ft) tall, and can be seen for miles around; it is accompanied by a second Buddha figure in brass – smaller, but still substantial at 12 metres (29 ft) tall. Buddhism is the religion of the majority here, but, as everywhere on the Malay Peninsula, there is a strong Muslim presence as well.

Phuket has been developing as a major tourist resort since the 1980s, and now tourism is the main source of income. But, being on the strategic shipping lanes between India and the Strait of Malacca, gateway to the Far East and Pacific, it has been receiving foreigners for centuries. The French were given a concession by the Thai king to mine tin here in the 17th century. The Burmese and British also tried to take possession of it. A famous landmark of the island is a bronze sculpture of two women warriors, Thao Thep Kasattri and Thao Si Sunthorn – the widow of a former governor and her sister who heroically led resistance to a Burmese invasion in 1785.

Every dream island is shadowed by its antithesis: nightmare. Phuket had become well established as one of the word's most desirable holiday resorts when calamity struck on 26 December 2004. The tsunami triggered by a submarine earthquake in the Indian Ocean swelled the sea into a low but

colossal series of waves that casually bulldozed their way into west coast of the island, killing at least 250 people and wrecking countless businesses.

But local people and the Thai government quickly recognized the necessity to rebuild and restore their tourist facilities as quickly as possible, and loyal visitors returned to show support. Today there is little visible sign of the disaster. And to back up reassurances about the extreme rarity of such occurrences, the Thai government has installed a series of satellite-linked early-warning buoys out in the Indian Ocean that will permit evacuation should another tsunami threaten.

The fish are so abundant that a fisherman casting a net in the shallows can hope to bring in a rewarding catch. Many aspects of traditional life still continue on Phuket, despite the dominance of tourism. The economy was once geared to tin-mining and rubber, and two-thirds of the island is still given over to agriculture.

A popular daytrip from Phuket takes visitors to the sugarloaf landscape of Phang-na to the north. James Bond Island is one feature, so-called because it appears in the 1974 film *The Man with the Golden Gun*.

BORNEO

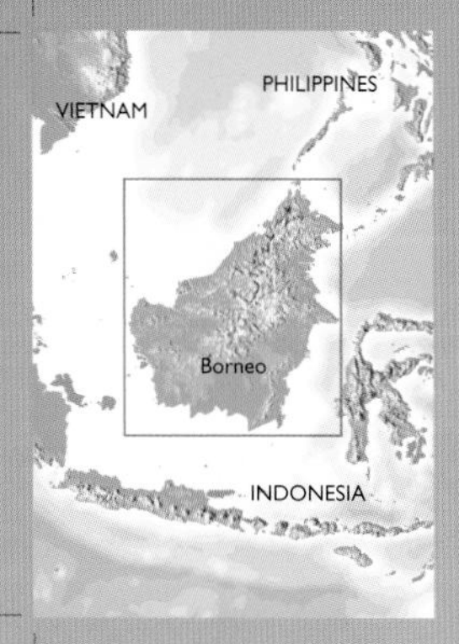

Latitude 3°21'N **Longitude** 117°35'E

Area 743,330 square kilometres (287,000 sq miles)

Status Shared by Indonesia, Malaysia and Brunei

Population 16,000,000

Official languages Malay, Indonesian

Currency Malaysian ringgit, Indonesian rupiah, Brunei dollar

Borneo is a vast island – the third largest in the world. Straddling the Equator, it is the Earth's hothouse, an experimental laboratory where nature was left to run riot for 130 million years in the world's oldest rainforests. The result is a dazzling profusion of life forms, each with its own carefully tailored strategy of survival, and constantly struggling to protect its niche. Some deliberately draw the eye with brilliant displays of colour, such as the huge Rajah Brooke's birdwing butterfly, and the crimson sunbirds and iridescent kingfishers. The mammals can be equally melodramatic, such as the

Mount Santubong, on the coast of Sarawak, is Borneo's Galápagos. Research here led the British naturalist Alfred Russel Wallace to develop his 'Sarawak Law', published in 1855 and paralleling Charles Darwin's Theory of Evolution.

clouded leopard, the armour-plated pangolin, and the endangered Sumatran rhinoceros and pygmy elephant.

Scale is sometimes outlandish: here a king cobra – the world's biggest venomous snake – can grow up to 6 metres (20 ft). There are giant poisonous centipedes, spiders and scorpions. Some insects of the forest floor are so beyond normal human imagination that they could be adopted for an alien filmset. New species of plants and animals are being discovered all the time – more than 350 in the past two decades alone.

TRAVELLER'S **TIPS**

Best time to go: Borneo is hot throughout the year, with average temperatures in the range of 23–31°C (73–88°F). It rains less in September–October.

Look out for: Islands off Sabah, such as Sibidan and those of the Tunku Abdul Rahman Marine Park, are world-renowned for snorkelling and scuba-diving.

Dos and don'ts: Do visit or stay in a longhouse, traditional dwelling of the Iban people of Sarawak. There is a Sarawak Cultural Village near Kuching.

Nature in Borneo does not do things by halves. It is home to the world's biggest flower, the *Raffelesia*, which blooms on parasitic vines on the rainforest floor and can measure 1 metre (3.3 ft) across. It is also called the 'corpse flower', because of its terrible stink, but this attracts the insects that it needs for pollination.

Humans have lived here for perhaps 40,000 years, and certainly 3,000 years in the case of the Dayaks – an umbrella term that covers some 200 indigenous peoples, speaking 150 languages. The bounty of the forests has also made Borneo a rich source of raw materials. In the past, Chinese traders came to these shores to collect hornbill ivory, rhinoceros horn, scented wood and edible birds' nests as well as gold. Hindu rajahs from the Majapahit Empire of Java controlled the coastal trade, and later sultans from Malaysia. European trading companies – German, British, Dutch – muscled in during the 19th century, and Sarawak was ruled by a series of British 'White Rajahs', starting with James Brooke, until 1946.

Borneo is now divided between three nations. The largest portion, in the south – occupying nearly three-quarters of the total – belongs to Indonesia, and is called Kalimantan. Most of the northern sector is owned by Malaysia, and is divided into two states: Sarawak to the south, and Sabah to the north. In between them is the tiny oil-rich state of Brunei.

Despite dramatic forest clearance undertaken by logging companies in pursuit of valuable tropical hardwoods, most of the island remains magnificently untamed. Rivers are still the only way to reach many parts of the interior. Intrepid travellers are drawn here to experience nature in its most ebullient and unforgiving mood – and to test their own skills of survival.

Borneo and Sumatra are now the only places where orangutans are found in the wild. Even there, their survival depends on the work of organizations such as the Borneo Orangutan Survival Foundation. This is a young male at the Lamandau Nature Reserve, Central Kalimantan.

NEW GUINEA

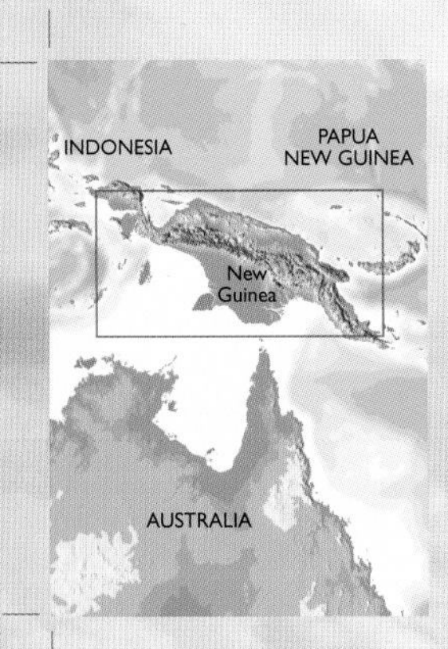

Latitude 5°20'S **Longitude** 141°36'E

Area 786,000 square kilometres (303,500 sq miles)

Status Shared by Papua New Guinea and Indonesia

Population 7,500,000

Capital Port Moresby (PNG), Jayapura (West Papua)

Official languages English, Tok Pisin, Hiri Motu (PNG); Indonesian (West Papua)

New Guinea has been called the last frontier of travel. In this vast island – the second largest in the world, after Greenland – there are dense tropical forests, active volcanoes and remote villages with Stone Age cultures. Travel is difficult in this under-explored landscape, and somewhat dangerous too. New Guinea has edge.

This is also a world where the geographical and zoological norms are turned on their heads. New Guinea lies just south of the Equator, yet has mountains so high that they have

glaciers on them. Its wildlife presents a catalogue of extremes: it has the world's largest butterfly, the Queen Alexandra's Birdwing, with a wingspan of 30 centimetres (12 in). It has birds of paradise, whose male performs a ritual mating dance in a blaze of dazzling colours. The male bower bird lures a female by creating gaudy homes of sticks decorated with anything that shines and glisters.

Like Australia to the south, New Guinea has marsupials: tree-kangaroos, the aggressive quoll or 'native cat', the

Raja Ampat ('Four Kings') is an archipelago of four islands and numerous islets, off the western tip of Indonesia's West Papua province. It has the greatest diversity of marine life in the world.

Shells, feathers and paint are all part of the extraordinary scope of invention that is garnered for the *sing-sing* events at the Highlands towns of Mount Hagen and Goroko. The ritualized clan warfare of the past is now played out in competitive festivals of traditional dance, which have become major events on the tourist calendar.

TRAVELLER'S **TIPS**

Best time to go: The best months are June–September. December–March is the (very) wet season, when many roads became impassable.

Look out for: Handmade crafts, such as masks, pottery, necklaces, baskets and string bags called *bilums* (large ones are strapped around the forehead).

Dos and don'ts: Be aware that both the sea and low-elevation freshwater lakes and rivers are home to saltwater crocodiles.

Ginetu is a tiny island in the Woodlark Group, some 300 kilometres (185 miles) off the southeastern tip of Papua. They form part of the Milne Bay Province, which alone contains more than 600 islands, of which 160 are inhabited – another example of New Guinea's mind-boggling statistics.

The female Queen Alexandra's Birdwing, the world's largest butterfly, is found only in one small area, around Popondetta in the Oro Province of southeastern Papua. The species is considered to be endangered. Queen Alexandra was the wife of King Edward VII, the ruling British monarch when the butterfly was named in 1907.

monkey-like cuscus and the spiny echidna, which looks like the product of a union between an anteater and a hedgehog, but is in fact a monotreme, an egg-laying mammal.

As in Australia, the people have been here for some 40,000 years. They compete with the wildlife in the sheer inventiveness of their adornment. For their festivals of dance and music – their *sing-sings* – they load their bodies with paint, beads, extravagant feathers, iridescent beetle shells, cow horns, dyed grass, human hair and improvised found objects such as beer caps and tin-can lids. They also echo the statistical extremes by speaking a total of nearly 1,000 languages between them. Only two forms of English-based pidgin serve as their unifying linguae francae.

Yet it would be wrong to think of New Guinea as a place untouched by the modern world. There are mines, coffee and copra plantations, and high-rise office buildings in the capital, Port Moresby. A man attending a *sing-sing* in full regalia up the Highlands town of Mount Hagen might arrive in a jeep with his golf clubs in the back. New Guinea receives a fair intake of tourists as well, numbering in tens of thousands. Many Australians come to hike along the Kokoda Trail, and to pay tribute to the bitter struggle waged by their countrymen during the Second World War to halt the Japanese advance on Port Moresby.

Other visitors come for the indigenous culture. They travel by boat up the serpentine Sepik River in the north – the longest river on the island – stopping at traditional villages where the *bigman* (chief) will show them the thatched *haus tambaran*, or spirit house, richly decorated with carved wooden sculptures.

Through the whims of colonial history, this vast island is divided in two by a near-straight border that slices right down the middle. To the east is Papua New Guinea, formerly an Australian-mandated territory that became independent in 1975; it also encompasses the substantial islands of the Bismarck Archipelago to the northeast. On the other side of the border is West Papua (previously called Irian Jaya); this half of the island was once a part of Dutch Indonesia, and is now governed by Indonesia.

Most travellers go to Papua New Guinea, but Indonesian Papua is open to visitors too, presenting perhaps an even greater and more exotic challenge. Take it on, and you will be rewarded by immense and untrammelled tracts of land. The Lorentz National Park, for instance, covers 25,056 square kilometres (9,674 sq miles) – an area bigger than Wales or New Hampshire – and includes Puncak Jaya (4,884 metres/16,024 ft), the highest mountain between the Himalayas and the Andes. The park is home to the dingiso, a tree-kangaroo first identified in 1995, but always known to the local Moni people who protected it as an ancestor. New Guinea remains an island of endless discovery.

BALI

Latitude 8°39'S **Longitude** 115°13'E

Area 5,633 square kilometres (2,175 sq miles)

Status Province of Indonesia

Population 3,891,400

Capital Denpasar

Official language Indonesian

Currency Indonesian rupiah

Every day tiny woven baskets bearing offerings – grains of rice, freshly picked flowers, betel nuts, incense sticks – are laid at the places where the gods and spirits reside: thresholds, gates, shrines, statues, old trees, cars, motor scooters, office desks. They are like physical prayers, little acts of worship calling for blessings and protection, poetic gestures to demonstrate respect for the sanctity of everything in our world.

Bali is imbued with spiritual devotion, expressed with the natural, almost casual sense of artistic beauty that seems to be

in the Balinese genes. For temple ceremonies, offerings of tall, colourful stacks of fruit and cakes arrive on the heads of women dressed in their finest clothes, to be massed among the open-air shrines. With some 20,000 temples in Bali, each with at least one festival a year, the calendar is packed. Prayers and devotions continue well into the night, accompanied by performances of dance, shadow puppets, theatre and the complex rhythms of the gamelan orchestras. Visitors to Bali will see the dances and hear the gamelan music, and be amazed at the intensely disciplined performances. They may think that these survive merely as tourist entertainment. They are wrong: the Balinese arts are for the gods; they are rolled out for the tourists only as a way of raising income, which can be invested in new gamelan instruments, costumes or some communal village project. And the very best performances are reserved for the temple ceremonies – utterly spellbinding moments beneath a full moon, when the gods are watching.

The Balinese may have good reason to thank the gods. This is a spectacularly beautiful island, where rich soil produces an

The tiered shrines of the Ulun Danu temple, thatched in *alang-alang* grass, are reflected in the water of Lake Bratan, in the hills of central Bali.

TRAVELLER'S **TIPS**

Best time to go: The dry season lasts from April to September, and May is considered the best month. The wet season lasts from October to March.

Look out for: Balinese craftworkers produce carved and painted wooden sculptures, batik and ikat woven cloth, paintings, silverware, jewellery and more.

Dos and don'ts: Do go to a temple or cremation ceremony (ask at your hotel). Visitors are welcome as long as they dress correctly and remain unobtrusive.

The Balinese have exceptional gifts for the arts, notably wood carving, a tradition that has been applied to temple decoration for centuries. Making imaginative use of gnarled pieces of timber is one aspect of their art, but sculptors are also highly flexible, mass-producing skilled – and often witty – painted pieces for the souvenir market.

abundance of flowering, fruit-bearing trees and bushes, and the rice paddies – watered by ancient irrigation systems – quilt the hillsides with emerald-green terraces. It is a landscape busy with visually ravishing scenes – a boy shepherding a line of ducks along the road to feed in the paddy fields; mossy, overgrown temples; distant views over the palms to the sacred volcano, Mount Agung, feathered by wisps of white cloud.

The Balinese appreciate this beauty too, just as they know their luck in inhabiting this unique island – which is a Hindu island among the 13,000 predominantly Muslim islands that make up Indonesia. Hinduism used to be the religion of neighbouring Java, brought from India in the 2nd century AD and adopted with the full pantheon of gods. But the Hindu rulers were put under increasing pressure by the more meritocratic Islam, until, in 1515, the last rajahs fled with their entourages to Bali. Here they lived in almost total isolation for nearly four centuries. Meanwhile all the neighbouring islands submitted to Dutch colonial rule. Finally, in 1906, when the Dutch could no longer tolerate this anomaly, they invaded Bali, provoking a ritual mass suicide by the royal families, who could not accept such humiliation.

Tourism is the mainstay of the Balinese economy, and the Balinese are determined to control development to preserve the unique charms which are the island's prime lure. When it opened its doors to mass tourism, Bali cunningly isolated the main development on Nusa Dua, a virtually uninhabited southern peninsula. They cannot permit their precious island to be spoiled: the gods would not be pleased.

A woman carries a basket of flowers, used as temple decorations and offerings, through the rice terraces of central Bali. The strong sense of community in Bali villages is said to derive from the complex patterns of sharing that are needed to irrigate the rice paddies effectively.

HOKKAIDO

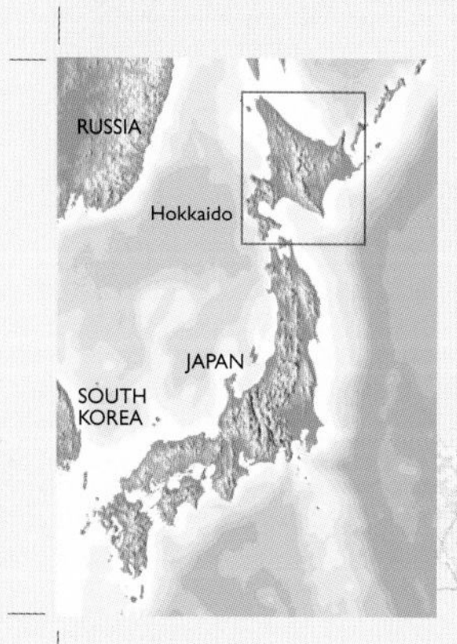

Latitude 43°0'N **Longitude** 142°0'E

Area 83,453 square kilometres (32,221 sq miles)

Status Prefecture of Japan

Population 5,507,500

Capital Sapporo

Official language Japanese

Currency Yen

White-water rafting, sea-kayaking, surfing, climbing, mountain-biking, skiing, snowboarding – Hokkaido, the most northerly of Japan's main islands, has developed rapidly in just the past few decades as a destination of choice for the Japanese, and for many international visitors, in search of adventure and activity holidays. The main city, Sapporo, came to world renown, of course, when it hosted the Winter Olympics in 1972. Now Niseko, Furano and Rusutsu attract skiers and snowboarders from around the world, and particularly from Australia – lured by the exceptional quality of the soft powder snow.

In other seasons, visitors come to admire the scenery – mountainous, forested and dotted with lakes. Much of Hokkaido's surface area is protected as parks, with seven national parks, a further five 'quasi-national' (which are administered by the prefecture) and twelve 'natural parks'. The dramatic contours of the landscape have been shaped by a long history of volcanic activity. There are five volcanoes classed as active: Mount Tarumae, which rises spectacularly beside Lake Shikotsu in the Shikotsu-Toya National Park, erupted ten times in the 20th century, the last time in 1982. By the same token, Hokkaido is regularly shaken by tremors and earthquakes that occur on average once a week. This

TRAVELLER'S **TIPS**

- **Best time to go:** For the Japanese, the high season is June–July, but spring and autumn are also pleasant. The main winter sports season is December–April.

Look out for: Hokkaido is celebrated for its beers. Just about every town has a microbrewery, and the Sapporo beer brand is internationally famous.

Dos and don'ts: Do go camping, but remember that it can be decidedly chilly at night in Hokkaido, even in the summer.

Autumn colours set ablaze the landscape of the Kamikawa subprefecture of central Hokkaido, near Asahikawa. Mixed forests are one of Hokkaido's distinctive features; conifers and broad-leaved trees coexist, forming some outstanding scenery.

activity also accounts for the numerous hot springs and thermal spas on the island.

Hokkaido lies 23 kilometres (14 miles) north of Honshu across the Tsugaru Strait. It is linked by the Seikan Tunnel, the world's longest and deepest rail tunnel, and also by ferry and air. Japanese tourists flock to Hokkaido to escape the summer heat and humidity that oppresses the other islands, especially during the rainy season of June and July. They also enjoy the sense of space. Honshu, Japan's largest island, is three times the size of Hokkaido, but has eighteen times its population.

Hokkaido remained completely undeveloped for centuries, inhabited mainly by the Ainu people, who also had a foothold across the sea in eastern Russia. When Russia developed Vladivostok in the 19th century and looked poised to expand into the Far East, the Japanese government took action to protect its claim to Hokkaido. The Russian island of Sakhalin, after all, lies only 50 kilometres (30 miles) to the north of Hokkaido. During the 1870s Japan invited American experts to Sapporo to develop agriculture and mining, and the population of the island soared from 58,000 to 240,000 in just one decade.

Agriculture remains a key pursuit, growing typically temperate-climate crops such as wheat, soya beans, and potatoes. Hokkaido has a dairy industry – and produces ice cream with distinctly unusual flavours, such as asparagus and

Whooper swans are a winter feature of Lake Kussharo. They gather in the warm outflow from the hot springs – which are also much appreciated by human bathers. Kussharo, in Akan National Park, eastern Hokkaido, is Japan's largest caldera lake.

squid ink. The city of Hakodate is renowned for its seafood – especially crab and squid. At the beginning of August each year, Hakodate celebrates a Port Festival, when thousands gather to perform the Ika-Odori, or Squid Dance.

As a result of its history, Hokkaido's cities are modern, and few buildings date back more than a century. Sapporo, the fifth largest city in Japan, is a bustling metropolis built on a grid and filled with traffic and neon signs. But, as if to make up for this geometric predictability, each February it hosts a Snow Festival – one of the biggest winter events in Japan. Two million visitors come to see a dream cityscape carved out of ice, including the world's historic monuments and fantasy palaces, along with hundreds of huge and intricate sculptures.

A boardwalk leads visitors through Jigokudani (Hell Valley), a tortured volcanic world, pungent with the smell of sulphur, and made more dramatic by nighttime illuminations. It lies above Noboribetsu Onsen, a village famous for its hot springs. Spa hotels here offer a range of baths; many promote the medicinal benefits of the mineral-rich water.

RYUKYU ISLANDS

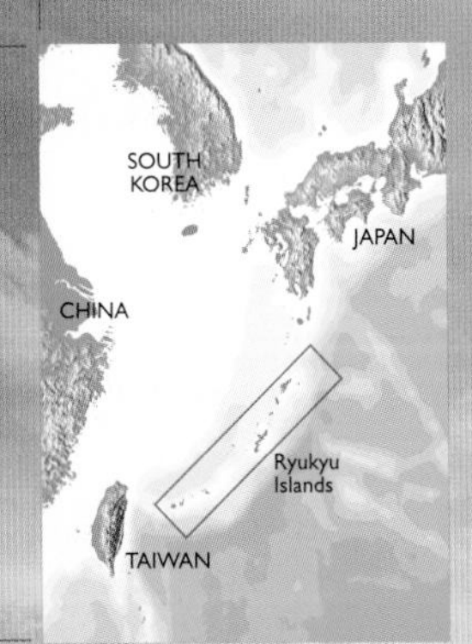

Latitude 26°19'N **Longitude** 127°44'E

Area 2,254 square kilometres (870 sq miles)

Status Islands of Japan

Population 1,320,000

Main settlement Naha (on Okinawa)

Official language Japanese

Currency Yen

The Ryukyu Islands stretch southwards in a long, arcing chain for 1,200 kilometres (750 miles), to a point just north of the Tropic of Cancer. Japanese visitors fly out in huge numbers to the larger islands, such as Ishigaki in the southernmost Yaeyama group, then take boats to the smaller islands. Here they find superb white-sand beaches, and some of the world's best diving, with coral reefs, turtles, hammerhead sharks and manta rays. On Iriomote they visit the National Park, one of Japan's great wildernesses, with rainforest, waterfalls and a chance of seeing the rare Iriomote wild cat. On Taketomi, they

Sunrise on Ishigaki Island will soon reveal turquoise seas and coral sands – the island dream offered by many of the 49 islands in the Okinawa-Ryukyu group, in the southern half of the chain.

TRAVELLER'S **TIPS**

Best time to go: Most agreeable are October–November and March–April (the 'Golden Week' of end April/early May gets very busy).

Look out for: Pairs of ceramic shisa lion-dogs, which guard entrances. One has an open mouth to catch good fortune, the other a closed mouth to keep it in.

Dos and don'ts: Beware of two venomous animals: on land, the habu viper, and, in the sea, the anbonia cone shell, which has a lethal stabbing appendage.

can visit a preserved Ryukyu village, with streets of sand and lush gardens walled with volcanic stone.

The people of the Ryukyu islands see themselves as different from the rest of the Japanese. They have a distinct language, their own music and cuisine, and a separate history. Their islands were dominated by China until an invasion by 2,500 samurai in 1609 brought them into the Japanese orbit. They then served two masters until the islands were formally annexed by Japan in 1879. The largest island of the chain is Okinawa, which has more centenarians per 100,000 people that anywhere else in the world. Diet, exercise, a lack of stress and a positive attitude are thought to be the winning combination. But maybe inhabiting a beautiful subtropical island also has something to do with it.

Cape Irizaki on the island of Yonaguni, at the base of the Ryukyu chain, marks the most westerly point in Japan, as this monument explains. Yonaguni is famous for the undersea Yonaguni Monument, frequented by divers: this stepped, stone construction seems too perfect to be natural, but if it is manmade, no one knows by whom or how.

HAINAN

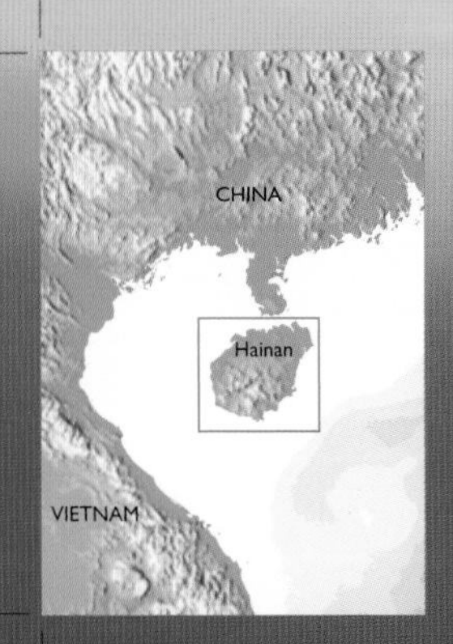

Latitude 19°12'N **Longitude** 109°42'E

Area 33,920 square kilometres (13,100 sq miles)

Status Province of the People's Republic of China

Population 8,640,700

Capital Haikou

Official language Mandarin Chinese

Currency Chinese yuan

Hainan's old traditions linger on in the coastal fishing villages and rural interior, and are now cherished for their nostalgic appeal.

Hainan is the most southerly point of China, lying well within the Tropics. With the South China Sea to its east, it hangs like a football sailing into the net formed by the Gulf of Tongking and the long coastline of Vietnam. Almost as big as Taiwan, it is a substantial body of land, rising in the centre to 1,867 metres (6,125 ft) at Wuzhi Mountain. Most of it is covered by tropical rainforest and agricultural land where rice, coconut palms, rubber trees, coffee and tea are grown, as well as tropical fruits such as pineapples and dragon fruit. But – most importantly – the perimeter is a chain of fine sandy beaches.

Just a quarter of a century ago, Hainan was a sleepy tropical island, little known to the outside world and little visited. Then, in 1987, as China began accelerating into a modern industrial giant, the government realized Hainan's potential as an international tourist resort. The island acquired the status of a Special Economic Zone, providing tax concessions and other benefits to encourage foreign investment. Hainan is now on the world tourism map, with 25 million visitors a year; most of these come from other parts China, but close on a million come from abroad. Wealthy Chinese from the north holiday here in January and February to escape the winter, as do many Russians. Hainan has year-round sunshine, and international visitors to China add Hainan to their itinerary as a place to relax after intense cultural tours on the mainland.

The Tianya Haijiao Tourist Zone, 25 kilometres (15 miles) west of Sanya, is tailor-made to deliver the tropical-beach dream. The giant, smooth rocks scattered at the water's edge are a famous landmark; two have been inscribed with the poetic meaning of the place name: 'The Edge of the Sky' and 'The End of the Sea'.

Sanya in particular has been the focus of development. Nicknamed 'China's Hawaii' – because it lies on the same latitude – it now offers a full range of resort-holiday facilities and attractions: windsurfing, scuba-diving, paragliding, jet

TRAVELLER'S **TIPS**

Best time to go: November to March. The summer rainy (and typhoon) season lasts from May to October; the north can be especially hot in July–August.

Look out for: Pearls: Hainan is famous for its cultured pearls. But to make a wise purchase, you need to have a sound prior knowledge of value and quality.

Dos and don'ts: Don't forget that Hainan is a malaria zone, so be sure to take the necessary precautions to protect yourself from infection.

skiing, golf. Links with the mainland have been steadily improving: there are airports at Sanya as well as at Haikou, the capital, in the far north of the island. Ferries cross the Qiongzhou Strait that separates Hainan from the mainland by a mere 30 kilometres (19 miles), including – since 2003 – a ferry which carries an entire train to Haikou, before an onward coastal journey to Sanya.

All this has been a sudden but not unwelcome change for the people of Hainan. For thousands of years this was the home of the Li people, living in their villages of thatched houses. Despite the proximity of the mainland, their island was mostly forgotten or sidelined by history. This was where criminals were sent, or disgraced officials exiled by emperor. After the Communists wrested control of the island from the Kuomintang (National Party) in 1950, it became a military and naval base – so close to Vietnam – and hence off-limits.

Tourist development is largely limited to pockets on the coasts, but a virtue is also being made of the unusually unspoilt interior, with new hotels promoting eco-tourism, and initiatives such as the Yanoda Rainforest Culture Tourism Zone and the Xinglong Tropical Botanical Garden in the Xinglong Hot Spring Tourist Zone. On Nanshan (Nan Mountain), near Sanya, visitors can admire ancient dragon trees, the Hainan *Dracaena*, some of which are said to be at least 6,000 years old, and seem in some way connected to the great longevity of local people there. A Chinese saying goes: 'May your age be as great as Nanshan.'

On the coast at Sanya, a glittering new statue of the Buddhist goddess Guanyin stands 108 metres (354 ft) tall – taller than the Statue of Liberty, with its plinth included. It seems an apt symbol of both the rapid rise of this newly fledged holiday island, and its determination to maintain a respectful bond with the past.

The annual Haikou Hot Air Balloon Festival lifts off from a park beneath the modern backdrop of the capital city. The event fits Haikou's reputation as a gentle, laid-back place, despite modern development.

EPSON
滩中心

TASMANIA

Latitude 42°0'S **Longitude** 147°0'E

Area 68,401 square kilometres (26,410 sq miles)

Status State of Australia

Population 507,600

Capital Hobart

Official language English

Currency Australian dollar

Tasmania is Australia's little secret. It hovers like an afterthought 240 kilometres (150 miles) to the south of southeastern Australia, and a 10-hour ferry journey from Melbourne. It is like no other part of Australia – a self-contained, triangular island, about the size of Sri Lanka, but with a cool, temperate climate and packed with the most extraordinary diversity of landscape. In acknowledgement of this, a full 40 per cent of the island is protected by national parks and nature reserves.

On the east coast there are beautiful bays with long, curving sandy beaches. By contrast, dolerite columns soar up from the sea to 300 metres (1,000 ft) in the Tasman National Park on the same coast. Across much of the island, the land rises rapidly from the coast, creating dramatic backdrops to the ports – most notably at Mount Wellington, which rears up behind Hobart, the capital nestled into the shelter of its bay in the southeast of the island. Nature is always close at hand. Launceston, the second city, located in the north, has the wild Cataract Gorge within a few minutes' walk of the city centre.

Most of the west of the island is mountainous, providing magnificent opportunities for hikers. One of the most famous walks is the Overland Track, running 65 kilometres (40 miles) through the Cradle Mountain-Lake St Clair National Park. Walkers are rewarded with landscapes of jagged, alpine mountain peaks, lakes, waterfalls and pristine woodland. In the Franklin-Gordon Wild Rivers National Park, close to Macquarie Harbour on the west coast, visitors can tour the wilderness on river cruises. Tasmania used to have a reputation for being an island primarily geared up for adventure and activity holidays – walking, kayaking, rafting, surfing, fishing,

Cradle Mountain, reflected in Dove Lake, is at the heart of the Cradle Mountain-Lake St Clair National Park in northwestern Tasmania.

TRAVELLER'S **TIPS**

Best time to go: Tasmania has four distinct seasons. Summer is from December to February. Spring and autumn have their rewards, but can be wet and foggy.

Look out for: The Mount Wellington Descent is a thrilling three-hour bike ride, dropping from 1,270 metres (4,167 ft) to sea level at the Hobart Waterfront.

Dos and don'ts: Do bring walking shoes or boots. You will do plenty of walking in Tasmania – in the towns as well as in the national parks and on the biking paths

The Tasmanian devil has earned its name from its sharp teeth and its haunting shriek. The size of a small dog, it is the largest carnivorous marsupial. Despite its ferocious reputation, it tends to be shy of humans. It has been adopted by Tasmania as the state emblem. As the result of a mysterious facial cancer, it has been listed as endangered.

The strange landscape of dolerite columns are a distinctive feature of Mount Wellington, which rises above the Derwent Estuary, and forms the backdrop to Hobart. With its summit at 1,271 metres (4,170 ft), it is frequently dusted with snow, even in summer.

scuba-diving among the kelp forests. These days, however, it has extended its appeal to include luxury designer-built boutique hotels and gourmet restaurants.

Nonetheless, Tasmania still retains a hard edge, which is enmeshed with its history. Although the Dutch navigator Abel Tasman came across the island in 1642, naming it Van Diemen's Land after the governor-general of the Dutch East India Company who had sent him there, it was not until 1803 that the British moved in and began to settle the island as a penal colony. It continued to take convicts well after Sydney had ceased to do so, with the result that many of the grander historic buildings on the island were built by convict labour. One of the most visited sites is Port Arthur, the best preserved convict settlement in Australia, and a moving testament to this era of Australian history.

Among the dramatic landscapes, there are plenty of reminders of the tough, proud lives of the early settlers: 19th-century port-side warehouses, farmsteads, fishing villages and mining towns. Hobart and Launceston are among Australia's oldest cities. Many of the place names suggest a desire to create Britain anew, in this similar, but fundamentally different world. The National Trust of Australia looks after Clarendon, a grand Georgian-style stately home near Launceston – not quite Britain, not quite Australia: distinctly Tasmanian.

TRAVELLER'S **TIPS**

Best time to go: April to October. Avoid January to March: this is the rainy season. The island is very busy during school holidays.

Look out for: The Fraser Island Great Walk: a trail of 90 kilometres (56 miles) through all the island's landscapes.

Dos and don'ts: Never feed the dingoes – or leave out any food or rubbish that they could eat. Offenders are liable to hefty on-the-spot fines.

FRASER ISLAND

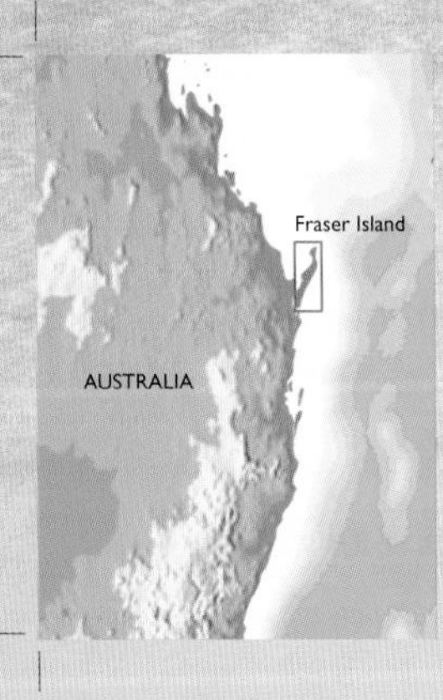

Latitude 25°13'S **Longitude** 153°8'E

Area 1,840 square kilometres (710 sq miles)

Status Island of Queensland, Australia

Population 360

Main settlement Eurong

Official language English

Currency Australian dollar

White silica sand accentuates the deep blues in the crystal-clear freshwater of Lake MacKenzie, a 'perched lake' to the west of the resort settlement of Eurong.

The world's largest sand island lies just off the Queensland coast, pointing a crooked finger northwards towards the Tropic of Capricorn and Great Barrier Reef. For some 750,000 years, currents sweeping up the coast have deposited sand here, creating an island that is 120 kilometres (75 miles) long, and just 24 kilometres (15 miles) wide. Its ancient dunes rise to 250 metres (820 ft).

The label 'sand island' might suggest a desert, but that is far from the case. With high rainfall, the soil is fertile enough to grow a thick coat of vegetation – eucalyptus woods, mangrove swamps, even rainforest – so from the air it would be hard to guess that this is a sand island at all. The only hints are patches of bare dunes, or 'sandblow', and the beaches that rim the island, including the enormous sandy beach that runs for almost the entire length of the east coast.

Flying over Fraser Island, you will see dozens of shallow lakes, like scattered sequins. These are 'perched lakes' – perched over a layer of fused, impermeable sand, capturing rainwater, their translucence often tinctured by tannins from decayed vegetation. The northern half of Fraser Island is protected as part of the Great Sandy National Park (which also covers areas in the adjacent mainland), and nature certainly is the island's great attraction. Visitors arrive by ferry from various points near Hervey Bay, to stay in resort hotels or camp, and to swim, surf and hike. Many tour the island in off-road 4x4s, including along the beaches at low tide, where the hard sand permits a maximum speed of 80 kilometres per hour (50 mph).

The wildlife includes many of the animals readily associated with Australia, such as kangaroos, wallabies, echidnas, possums, geckos and goannas, as well as the occasional saltwater crocodile and 19 types of snake. From August to October, you might catch a glimpse of the humpback whales that migrate past the coast, and there are more than 350 species of bird, including kookaburras, honeyeaters and cockatoos. But perhaps the most famous, or notorious, resident is the dingo.

Isolated from the mainland, Fraser Island is thought to have the purest strain of dingoes in Australia – that is to say, they are not mixed with domestic dogs. Management of the dingoes has become a major issue, and visitors cannot bring dogs to the island; they are not permitted to feed the dingoes, and may be fined for doing so. Dingoes are scavengers, and will invade campsites and the small settlements on the island if they can circumvent the high fences that have been erected to keep them out. But the

Tracks, originally forged by loggers for their bullock carts, run through pockets of rainforest in the interior. Logging camps were set up in the 19th century to exploit kauri pines and gums (eucalyptus) and particularly the tall, straight satinays, or Fraser Island Turpentines. The forests have flourished again since logging ceased in 1991.

population is declining, and there are signs that the dingoes are under stress, which may also be the cause of a number of incidents in which dingoes have attacked humans.

Fraser Island takes its name from a tragic incident involving a Scottish sea captain, James Fraser, and his wife Eliza. They survived a shipwreck in 1836, and managed to reach the island in a longboat, only to be captured by the local Aborigines, the Butchulla. According to Eliza's account, they were cruelly mistreated and Captain Fraser died before she was rescued, six weeks after the shipwreck, by an escaped convict. Her story became a sensation, and Eliza profited unscrupulously from it. The details may well have been embroidered for dramatic effect, and remain the subject of controversy to this day.

The Butchulla people had lived on the island for some 5,000 years, before they were decimated by the disease and degradation that accompanied white settlement – mainly for logging – in the late 19th century. The surviving population was relocated in Queensland in 1904. Their name for the island was K'gari, meaning 'paradise', and today – with the island's natural wonders now under careful protection – it is easy to see why.

Sunset over Lake Boomanjin accentuates the golden tint of its water, caused by tannins leached from the surrounding plants – yet this freshwater lake is considered to be very pure.

INDEX

N

O P

R

S

T

U V

W Y Z

ACKNOWLEDGEMENTS

Quercus Publishing Plc
55 Baker Street
7th Floor, South Block
London
W1U 8EW

First published in 2012

Text by Antony Mason

A catalogue record of this book is available from the British Library

UK and associated territories:
ISBN 978-1-78087-156-1

Printed and bound in China

10 9 8 7 6 5 4 3 2 1

Designed and edited by
Therefore Publishing Limited

Maps by Map Graphics Limited

Picture Credits

2–3 Ian West/Getty Images; 8–9 Peter Landon/Getty Images; 10–11 Image Source/Getty Images; 11 Sarah M. Golonka/Getty Images; 12–13 David C Tomlinson/Getty Images; 14–15 Digitaler Lumpensammler/Getty Images; 15 Daryl Benson/Getty Images; 16–17 Philipp Klinger/ Getty Images;18–19 Travel Ink/Getty Images; 19 Slow Images/Getty Images; 20–21 Felix St. Clair Renard/Getty Images; 21 Sisse Brimberg/Cotton Coulson/Keenpress/ Getty Images; 22–23 Johan Odmann/Getty Images; 23 Mattias Nilsson/Getty Images; 24–25 Daryl Benson/ Getty Images; 26–27 Emmanuel Coupe/Getty Images; 28–29 gmsphotography/Getty Images; 30–31 Andrew Holt/Getty Images; 32–33 The Edge Digital Photography/ Getty Images; 34–35 Chris Ladd/Getty Images; 36–37 Ingmar Wesemann/Getty Images; 37 Getty Images; 38–39 Hauke Dressler/Getty Images; 40–41 Dorling Kindersley/Getty Images; 42–43 Mark Banks/Getty Images; 44–45 Michele Falzone/Getty Images; 46–47 Darrell Gulin/Getty Images; 47 DEA/G. DAGLI ORTI/ Getty Images; 48 Loraine Wilson/Getty Images; 48–49 Andrea Pistolesi/Getty Images; 50–51 Arctic-Images/ Getty Images; 52 Philippe Bourseiller/Getty Images; 52–53 Ernst Haas/Getty Images; 54–55 Harri's Photography/Getty Images; 56 Jens Kuhfs/Getty Images; 56–57 Svein Nordrum/Getty Images; 58–59 Rinie Van Meurs/ Foto Natura/Getty Images; 59 Paul Oomen/ Getty Images; 60–61 Daisy Gilardini/Getty Images; 62–63 Simeone Huber/Getty Images; 64–65 Gavin Hellier/Getty Images; 65 Panoramic Images/Getty Images; 66–67 AFP/Getty Images; 68 Tony Souter/Getty Images; 68–69 Image Source/Getty Images; 70–71 Sylvain Sonnet/ Getty Images; 72 Bruno Barbier/Getty Images; 72–73 Anger O./Getty Images; 74–75 Slow Images/ Getty Images; 76–77 Chris Parker/Getty Images; 78–79 Herve Hughes/Getty Images; 80–81 Getty Images; 81 Maremagnum/Getty Images; 82–83 Wayne Lynch/ Getty Images; 84 Joseph Van Os/Getty Images; 84–85 Jim Brandenburg/Getty Images; 86–87 Lee Frost/Getty Images; 87 DEA/V. GIANNELLA/Getty Images; 88–89 Donald Nausbaum/Getty Images; 89 Design Pics/David DuChemin/Getty Images; 90–91 Andre Gallant/Getty Images; 92–93 altrendo travel/Getty Images; 94–95 David Sanger/Getty Images; 96 Sylvain Sonnet/ Getty Images; 96–97 Jake Rajs/Getty Images; 98–99 Jean-Pierre Pieuchot/Getty Images; 100 Andre Gallant/ Getty Images; 100–101 Walter Bibikow/Getty Images; 102–103 Michele Falzone/Getty Images; 104–105 Danita Delimont/Getty Images; 106 Kevin Moloney/Getty Images; 106–107 Purestock/Getty Images; 108–109 Jon Hicks/Getty Images; 110–111 J. Greg Hinson, MD, www. ackdoc.com/Getty Images; 111 Panoramic Images/Getty Images; 112–113 Harvey Lloyd/Getty Images; 114–115 Jose Azel/Getty Images; 116–117 Tai Power Seeff/Getty Images; 118 John Burcham/Getty Images; 118–119 Sean Davey/Getty Images; 120–121 Louise Murray/Getty Images; 122–123 Gavin Hellier/Getty Images; 124–125 Grafissimo/Getty Images; 126–127 Art Wolfe/Getty Images; 128 Danita Delimont/Getty Images; 128–129 Luciano Candisani/Getty Images; 130–131 M.M. Sweet/ Getty Images; 132–133 Photodisc/Getty Images; 133 Diane Cook and Len Jenshel/Getty Images; 134–135 Diane Cook and Len Jenshel/Getty Images; 136–137 Image Source/Getty Images; 137 Image Source/ Getty Images; 138–139 altrendo travel/Getty Images; 140–141 Thomas Pickard/Getty Images; 142–143 Caroline von Tuempling/Getty Images; 143 Reinhard Dirscherl/Visuals Unlimited, Inc./Getty Images; 144–145 altrendo travel/Getty Images; 146–147 Berndt Fischer/Getty Images; 147 Martin Harvey/Getty Images; 148–149 AZAM Jean-Paul/Getty Images; 149 Michael Runkel/Getty Images; 150–151 Federica Grassi/Getty Images; 152–153 Thomas Marent/Getty Images; 154 Thomas Marent/Getty Images; 154–155 Keren Su/ Getty Images; 156–157 tbradford/Getty Images; 158–159 Claudia Uribe/Getty Images; 159 vladimir zakharov/Getty Images; 160–161 Tim Laman/Getty Images; 162–163 Anup Shah/Getty Images; 163 Gerry Ellis/Getty Images; 164–165 Danita Delimont/Getty Images; 165 Tim Graham/Getty Images; 166 Le-Dung Ly/Getty Images; 166–167 Danita Delimont/Getty Images; 168–169 Carlina Teteris/Getty Images; 170 Martin Westlake/ Getty Images; 170–171 Martin Puddy/Getty Images; 172–173 travel Photo/a.collectionRF/Getty Images; 174–175 Ben Cranke/ Getty Images; 175 Masami Goto/Sebun Photo/Getty Images; 176–177 MIXA/Getty Images; 177 HIROAKI OTSUBO/SEBUN PHOTO/Getty Images; 178–179 IMAGEMORE Co.,Ltd./Getty Images; 179 Eurasia/ Getty Images; 180–181 Getty Images; 182–183 John White Photos/Getty Images; 184 Steve Turner/Getty Images; 184–185 Grant Dixon/Getty Images; 186–187 Jason Edwards/Getty Images; 188 Andrew Watson/Getty Images; 188–189 Natphotos/Getty Images

Title page: The sun rises behind a chapel built on an offshore rock formation at Georgioupolis, Crete.